Brilliant LED Projects

20 Electronic Designs for Artists, Hobbyists, and Experimenters

Nick Dossis

New York   Chicago   San Francisco   Lisbon   London   Madrid
Mexico City   Milan   New Delhi   San Juan   Seoul
Singapore   Sydney   Toronto

Cataloging-in-Publication Data is on file with the Library of Congress

McGraw-Hill books are available at special quantity discounts to use as premiums and sales promotions, or for use in corporate training programs. To contact a representative, please e-mail us at bulksales@mcgraw-hill.com.

Brilliant LED Projects: 20 Electronic Designs for Artists, Hobbyists, and Experimenters

Copyright © 2012 by The McGraw-Hill Companies. All rights reserved. Printed in the United States of America. Except as permitted under the Copyright Act of 1976, no part of this publication may be reproduced or distributed in any form or by any means, or stored in a database or retrieval system, without the prior written permission of publisher, with the exception that the program listings may be entered, stored, and executed in a computer system, but they may not be reproduced for publication.

McGraw-Hill, the McGraw-Hill Publishing logo, TAB™, and related trade dress are trademarks or registered trademarks of The McGraw-Hill Companies and/or its affiliates in the United States and other countries and may not be used without written permission. All other trademarks are the property of their respective owners. The McGraw-Hill Companies is not associated with any product or vendor mentioned in this book.

PIC® and PICkit™ are trademarks of Microchip Technology Inc.

1 2 3 4 5 6 7 8 9 0   QDB QDB   1 0 9 8 7 6 5 4 3 2

ISBN   978-0-07-177822-0
MHID   0-07-177822-5

Sponsoring Editor Roger Stewart	**Indexer** Jack Lewis
Editorial Supervisor Janet Walden	**Production Supervisor** George Anderson
Project Editor Patricia Wallenburg	**Composition** TypeWriting
Acquisitions Coordinator Molly Wyand	**Art Director, Cover** Jeff Weeks
Copy Editor Bill McManus	**Cover Designer** Pehrsson Design
Proofreader Paul Tyler	

Information has been obtained by McGraw-Hill from sources believed to be reliable. However, because of the possibility of human or mechanical error by our sources, McGraw-Hill, or others, McGraw-Hill does not guarantee the accuracy, adequacy, or completeness of any information and is not responsible for any errors or omissions or the results obtained from the use of such information.

*I dedicate this book to my late Grandfather Jack,
who was a big inspiration to me
and led me on the path to discovering electronics.*

About the Author

Nick Dossis lives in England and holds a Higher National Certificate in Electronic/Electrical Engineering. Several of his electronic projects have been published in *Everyday Practical Electronics (EPE) Magazine*, a popular monthly publication for electronics hobbyists. Nick has been playing around with electronics for most of his life, originally encouraged by his grandfather, who bought him his first crystal radio set when he was about seven years old. He continues to design electronic circuits and projects in his spare time.

Contents

	Acknowledgments	xi
	Introduction	xii
1	**Read This Before You Start Any Projects**	**1**
	Working with Stripboard	1
	How to Build Circuits on Stripboard	2
	Soldering Tips and Techniques	8
	Choosing Soldering Equipment	8
	Practicing Your Soldering Techniques	9
	Antistatic Precautions	12
	Working with Breadboard	13
	Programming PIC Microcontrollers	13
	Resistor Color Codes	15
	Circuit Diagrams and Stripboard Layouts	15

PART ONE   Illumination and Flasher Projects

2	**Basic LED Circuits: How to Make an LED Flashlight**	**21**
	What Are Light-Emitting Diodes?	21
	How to Illuminate an LED	22
	Illuminating Multiple LEDs	25
	Project 1 LED Flashlight	25
	Parts List	27
	How to Make the Flashlight	27
3	**An Alternative Way to Power an LED: "Green" Pocket LED Flashlight**	**31**
	Capacitors	31
	Using a Capacitor to Power an LED	32
	Project 2 "Green" Pocket LED Flashlight	34
	Safety Features	35
	How the Circuit Works	35
	Parts List	35
	Stripboard Layout	36

How to Build and Test the Board 36
Additional Safety Features 39
Possible Circuit Modifications 40

4 Building a Clock Generator: Basic Single-LED Flasher 41

The 555 Timer 42
   555 Timer Variants. 42
   555 Astable Timing Formulas 42
   Example 555 Astable Timing Calculations 43
   Source Current and Sink Current 44
Project 3 Basic Single-LED Flasher 45
   How the Circuit Works 45
   Parts List 45
   Stripboard Layout 46
   How to Build and Test the Board 46
   Experiments to Try on Your Own 49

5 Bringing a 555 Timer to Life: LED Bike Flasher 51

Project 4 LED Bike Flasher. 51
   How the Rear LED Flasher Circuit Works 52
   Parts List 53
   Stripboard Layout 54
   How to Build the Board 55
   Front LED Flasher. 56
   Enclosure for the LED Bike Flasher. 57
   Experimenting to Reduce the Current Consumption 59
   Alternative Circuit. 60

6 Exploring Multicolor LEDs: Color-Changing Light Box 61

Multicolor LEDs 61
   Bicolor LEDs. 61
   Tricolor LEDs 62
   Red, Green, and Blue (RGB) LEDs 62
   Symbols for Multicolor LEDs. 63
Project 5 Color-Changing Light Box. 63
   How the Circuit Works 63
   Parts List 65
   Stripboard Layout 66
   How to Build and Test the Board 66
   Finding an Enclosure. 68
   Light Me Up 70
   Possible Circuit Modifications 70

7 Using Seven-Segment Displays: Mini Digital Display Scoreboard 71

Seven-Segment Displays 71
Project 6 Mini Digital Display Scoreboard 73

How the Circuit Works	73
Parts List	75
Stripboard Layout	77
How to Build and Test the Board	77
Mounting the Board in an Enclosure	80
Future Modifications	82

PART TWO   Sequencer Projects

8 Introducing the 4017 Decade Counter: Experimental LED Sequencer Circuit 85

The 74HC Range of ICs	85
Project 7 Experimental LED Sequencer Circuit	86
How the Circuit Works	87
Parts List	88
Stripboard Layout	88
How to Build and Test the Board	90
Time to Experiment	92
Further Modifications	95

9 Driving Multiple LEDs from a Single IC Output: Color-Changing Disco Lights 97

Project 8 Color-Changing Disco Light	97
How the Circuit Works	98
Parts List	98
Deciding Which LED Enclosure to Use	101
Stripboard Layout	101
How to Build and Test the Board	102
How to Build the LED Display	104
Time to Disco	110

10 LED Binary Ripple Counter 111

The 4060 and 74HC4060 Binary Ripple Counters	112
Project 9 LED Binary Ripple Counter	112
How the Circuit Works	112
Parts List	113
Stripboard Layout	114
How to Build and Test the Board	114
Counting in Binary	116

11 Flickering LED Candle 119

Project 10 Flickering LED Candle	119
How the Circuit Works	119
Parts List	121
Stripboard Layout	122
How to Build and Test the Board	122
The LED Candle Enclosure	124

Experimenting with the Circuit................................... 124
Alternative IC?... 125

12 Introducing the PIC16F628-04/P Microcontroller: LED Scanner.. 127
The PIC16F628-04/P Microcontroller................................ 127
Project 11 The LED Scanner... 128
How the Circuit Works.. 128
Parts List... 130
Stripboard Layout.. 131
How to Build and Test the Board.................................. 131
The PIC Microcontroller Program.................................. 132

13 LED Light Sword.. 139
Project 12 The LED Light Sword..................................... 139
How the Circuit Works.. 139
Parts List... 141
How to Make the Enclosure.. 142
Stripboard Layout.. 144
How to Build and Test the Board.................................. 144
How to Make the LED String....................................... 145
Putting It All Together.. 146
The PIC Microcontroller Program.................................. 148
The Final Tests.. 149
Time to Play... 149

14 A Manually Operated Sequencer: Invisible Secret Code Display.................................. 151
Project 13 The Invisible Secret Code Display....................... 151
How the Circuit Works.. 152
Parts List... 154
Stripboard Layout.. 154
Preparing the Enclosure.. 155
How to Build and Test the Board.................................. 156
Putting It All Together.. 158
Send Me a Message.. 158
Future Modifications... 160

PART THREE POV Projects

15 Basic LED Matrix and POV Concepts: How to Build a Three-Digit Counter............................. 163
Persistence of Vision (POV)....................................... 163
LED Multiplexing Circuit Principles............................... 164
Project 14 Building the Three-Digit Counter....................... 164
How the Circuit Works.. 164
The Display Codes.. 166

	Parts List	167
	Stripboard Layout	168
	How to Build and Test the Board	168
	The PIC Microcontroller Program	170
	POV in Action	171
	Possible Enhancements	172
16	**A Multicolor POV LED Circuit: Backpack Illuminator**	**173**
	Project 15 Backpack Illuminator	173
	How the Circuit Works	174
	Parts List	176
	Stripboard Layout	177
	How to Build the Driver Board	177
	How to Build the Display Board	178
	Completing the Enclosure	180
	Test the Boards	180
	The PIC Microcontroller Program	181
	Testing Time	184
	Incorporating the Display into Fabric	184
	Other Ideas	186
17	**Using a Dot-Matrix Display to Show a Waveform: Digital Oscilloscope Screen**	**187**
	Project 16 Digital Oscilloscope Screen	187
	How the Circuit Works	188
	Parts List	191
	Stripboard Layout	192
	How to Build and Test the Board	192
	The PIC Microcontroller Program	195
	Give Me a Wave	196
	Other Ideas	199
18	**Light-Dependent LEDs: Experimental Low-Res Shadow Camera**	**201**
	Project 17 Experimental Low-Res Shadow Camera	201
	How the Circuit Works	202
	Parts List	203
	Stripboard Layout	204
	How to Build and Test the Board	204
	The PIC Microcontroller Program	207
	Mount the Board	208
	Seeing a Shadow	210
	Further Ideas	210
19	**Creating a POV Effect in Mid-Air: Groovy Light Stick**	**211**
	Project 18 Groovy Light Stick	211
	How the Circuit Works	211

	Parts List	212
	Stripboard Layout	213
	How to Build the Board	213
	Build the LED Display	214
	Complete the Enclosure	215
	Test the Board	217
	The PIC Microcontroller Program	218
	Making It Move	220
	Further Modifications	222

20 Showing Numbers on a Dot-Matrix Display: Dot-Matrix Counter . . . 223

Project 19 Dot-Matrix Counter . . . 223
   How the Circuit Works . . . 224
   Parts List . . . 225
   Stripboard Layout . . . 226
   How to Build and Test the Board . . . 227
   The PIC Microcontroller Program . . . 228
   Time to Count . . . 229
   Enclosure and Alternative Uses . . . 230

21 Creating Animations and Scrolling Text on a Dot-Matrix Display: Moving Message Destiny Predictor . . . 231

Project 20 Moving Message Destiny Predictor . . . 232
   How the Circuit Works . . . 232
   Parts List . . . 233
   Stripboard Layout . . . 234
   How to Build and Test the Board . . . 234
   The PIC Microcontroller Program . . . 236
   Construction Details . . . 242
   The Display Card . . . 243
   Discover Your Destiny! . . . 245
   Possible Program Modifications . . . 247
A Final Word from Me . . . 247

Appendix: Useful Resources . . . 249

Electronic Components Suppliers . . . 249
PIC Microcontroller Reference Books . . . 250
Electronics Hobbyist Magazines . . . 250
PIC Microcontrollers . . . 250
LochMaster 4.0 Stripboard Software . . . 250
Brilliant LED Projects . . . 250

Index . . . 251

Acknowledgments

There are a few people who I want to thank for helping me with this book. First of all, a big thank you to my family, who have supported me while I wrote this book and who have had to put up with my electronics tinkering over the years. My loving wife, Elissa Dossis, helped me with the fabric aspect of the backpack project, and you can find her own fabric creations at www.notjusthandbags.co.uk. A lot of the close-up photography in the book was expertly taken by Jasmine Dossis, and Georgia Dossis helped me to test out each of the projects once they were built. I'd also like to thank Paul Dossis, who designed the schematic symbols that I used to create the circuit diagrams in each of the projects.

Finally, I'd like to thank the editorial staff at McGraw-Hill, headed up by Roger Stewart, who gave me the opportunity to write this book and who have been very supportive during the whole writing process.

Introduction

I can remember that day in the 1970s as though it was yesterday. I was about seven or eight years old and staying with my grandparents in Liverpool, England, for a few days during the holidays. My late grandfather liked to collect things and always seemed to have some gadget or gizmo that he kept hidden away in his wardrobe. For example, I can remember a glow-in-the-dark pocket compass, a camera, and a red LED display calculator, to name a few. He liked to surprise me whenever I visited, and enjoyed seeing my eyes light up whenever he disappeared upstairs and brought down something new for me to have a look at.

This time he surprised me with a present; it was a crystal radio kit. The kit was fairly basic, composed of a rectangular plastic tray with a cardboard liner that had all of the components required to build a basic radio set. The components could be connected together by inserting interconnecting wires into the small springs that were connected to each component. After my initial excitement, I set to work building the radio, and within an hour I was listening to many faint radio stations through the crystal earpiece. I listened intently to some UK radio stations and even to a few others from further afield that were broadcasting in foreign languages. I couldn't understand the foreign speakers, but this did not matter to me—I had managed to build my first electronic project, and I loved it. Wanting to make the signals stronger, I soon discovered that I could get better reception if I connected the earth cable to a copper water pipe. I enjoyed building the kit, and from that day on my grandfather encouraged me to experiment with electronics. He regularly bought me *Everyday Practical Electronics*, a hobbyist magazine, for inspiration. I was soon hooked, and electronics became not only my hobby but a big part of my life, and it still is to this day.

I don't know what it is, but there is nothing like connecting together a handful of components and watching them come alive in some way, whether that be making a noise or flashing lights; it never becomes tiring to me. In fact, some of the easiest electronic circuits I learned how to build in my early days of experimenting were flashing-light projects, which not only were visually pleasing but also provided me with real satisfaction.

As its title suggests, this book contains a collection of electronic circuit projects that utilize the common light-emitting diode (LED) at their core. I really enjoyed putting these projects together and writing this book, and I really hope that you enjoy reading it and building the projects. If these pages inspire other hobbyists, artists, and experimenters to go on to make their own electronic designs in the future, then I will have helped to pass on my grandfather's inspiration to others.

What's the Book About?

This book contains a mixture of interesting electronic project circuits utilizing a broad selection of LED components, including standard single-color, tricolor, RBG, infrared, seven-segment, bar-graph, and dot-matrix displays.

The book is split into three main project sections:

- Part One: Illumination and Flasher Projects (basic LED projects)

- Part Two: Sequencer Projects (projects incorporating LEDs that illuminate in a particular sequence)

- Part Three: POV Projects (projects exploiting persistence of vision effects)

The projects use a variety of digital integrated circuits to achieve the desired results. You will learn how to work with CMOS 4000–range integrated circuits, 555 timers, bar-graph drivers, and the PIC16F628 PIC Microcontroller. I recommend that you read Chapter 1 first because it provides useful tips and recommendations that will be invaluable to you as you work through the projects in the book. Each subsequent chapter assumes that you have read and understood the details outlined in Chapter 1 and illuminates various circuit concepts and techniques and leads you through the construction of a project. The idea is for you to learn about various electronic building blocks and see how they fit into each project. This will teach you how the circuits work and enable you to go on to use them in your own project ideas.

Who Is the Book Aimed At?

This book is aimed at anyone who is interested in electronics and visual art. I have attempted to write this book in such a way that it appeals to electronics hobbyists and experimenters of all abilities, although it does assume that you have some knowledge of electronics and that you are able to read (or learn how to read) schematic diagrams. I would recommend that beginners read through the chapters and follow the projects sequentially, because the format of the book takes you on a journey that starts with basic LED circuits and then leads you through increasingly more complex projects. More experienced readers may choose to read about and build the projects in whichever order they prefer.

Each project contains a list of project specifications, a description of how the circuit works, a circuit diagram, a list of components required to build the circuit, a description and visual of the stripboard layout, an explanation of how to build and test the stripboard, and, where relevant, details about the PIC Microcontroller program. Readers can also download the programmable code that I have written for each of the projects that use a PIC Microcontroller from www.mhprofessional.com/computingdownload. I have also outlined my thought processes behind each design and how I arrived at the end result. Some of the projects are shown complete in enclosures while others rely on you to use your own imagination and choose your own method of mounting the circuits. The projects utilize a mixture of digital circuit techniques, and some of the projects demonstrate that sometimes more than one way to create a circuit design exists.

What Equipment Do I Need?

There are some basic pieces of equipment that you need to have to build the projects in this book, which are in addition to the components outlined in each project (for more information about obtaining these components, see the Appendix). The following is a list of the basic equipment required to get you started; if you decide to mount your circuits in elaborate enclosures, then you will probably require other tools and equipment. Chapter 1 discusses in more detail many of the items in this list.

- Soldering iron and stand, fume extractor, solder, and a desoldering tool
- 1/8" (3mm) drill bit or spot face cutter
- Stripboard polishing block or fine-grade sandpaper
- Safety glasses or goggles (for soldering, cutting, and drilling)
- Small hacksaw
- Wire cutters
- Antistatic mat and wrist strap
- Variable-speed drill and a selection of drill bits
- Digital multimeter with a minimum of voltage, current, and resistance settings
- PIC Microcontroller programmer and a personal computer (these are not required for all projects)

> **NOTE**
> Each constructional project has been extensively tested as part of writing this book; however, the author cannot guarantee the long-term performance, or accept legal responsibility for, the results of building these projects. The reader builds the projects outlined in this book at their own risk.

CHAPTER 1

Read This Before You Start Any Projects

I know that you are itching to get started on your first project, but I strongly recommend that you first read through this chapter because it contains some important tips, recommendations, and resources that will be useful to you as you work through the projects in this book. This chapter outlines some of the methods that I have adopted over many years of building electronic projects because they work well for me. I hope that you find these tips and techniques to be useful, and that you'll also discover other methods and skills that work well for you personally as you gain experience building your own projects.

Working with Stripboard

One frustration that I have had over the years when reading electronic project books is that sometimes the reader is shown a circuit diagram and given an explanation of how the circuit works, but is then left to decode the circuit and either build it on breadboard or make their own printed circuit board (PCB). Some readers may prefer this method, and you may decide to use the circuit diagrams in this book to produce your own PCBs. However, I have decided to also include instructions and photographs that show you how to build the circuits on stripboard. Using stripboard is an alternative method of building permanent electronic circuits that does not require you to make (or buy) your own bespoke PCBs. It's also an ideal method of building prototype circuits.

Stripboard is a copper-clad board that has a number of copper strips running along the width of the board and contains predrilled holes positioned at a pitch of 0.1" (2.54mm), which matches the format of standard through-hole electronic components (see Figure 1-1). The copper strips allow you to join the components together to create a circuit.

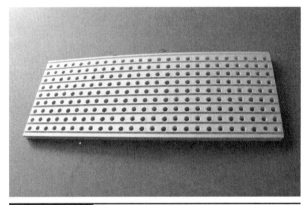

Figure 1-1 The copper side of a piece of stripboard

Stripboard is available in various sizes and strip and hole configurations, such as the two examples shown in Figure 1-2 (which are used in this book), and is also easy to cut and modify to suit your own requirements. You can cut the stripboard to size by using a small hacksaw or by carefully scoring the board a few times and then breaking the board over a piece of wood or a hard surface.

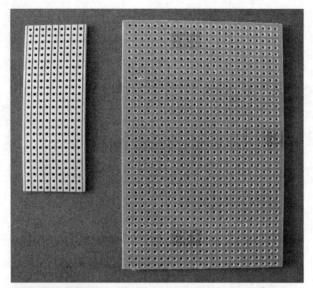

Figure 1-2 The two sizes of stripboard used in this book

If you decide to make your circuits on stripboard, you will need to know how to solder. If you don't feel confident in your soldering skills, then I recommend that you use breadboard to mount your circuits (see "Working with Breadboard" later in this chapter). Some of the experimental circuits in this book show you how to use breadboard.

To ensure that the final stripboard circuit design is compact, you will need to cut some of the tracks, as shown in Figure 1-3 (and as will be shown clearly in the stripboard layout images in this book). I also like to mark the noncopper side of the stripboard with a black marker to show where the tracks are cut on the copper side, which makes it easier to locate the position of the components before soldering them in place. It is very easy to cut the tracks, and I prefer to use a small 1/8" (3mm) drill bit fitted into a small wooden handle. Turning the drill bit carefully by hand is sufficient to remove the copper and create a gap in the track. There are also spot face cutting tools on the market that are designed to produce the same result.

I personally enjoy converting schematic diagrams into stripboard layouts. There are various methods of doing this. The simplest method is to use a piece of graph paper and a pen. There are also dedicated stripboard layout software packages available. I like to use LochMaster 4.0 software to create the stripboard layouts, which is the software I've used to create the stripboard layouts in this book. LochMaster 4.0 has a suite of components that you can pick and place on the stripboard layout on the screen, which makes layout adjustments easy to perform. Details of how to obtain this software are provided in the appendix of this book. You don't need to convert the schematic diagrams for the projects in this book to stripboard layouts because I have done this for you.

How to Build Circuits on Stripboard

You can use the following general procedure for any projects that require building circuits on stripboard. You should refer to this section when you build circuits in each of the projects in this book. As previously mentioned, you also need to be competent in soldering to build the projects on stripboard. Before you start to solder, be sure to read the nearby sidebar as well as the section "Soldering Tips and Techniques" later in this chapter.

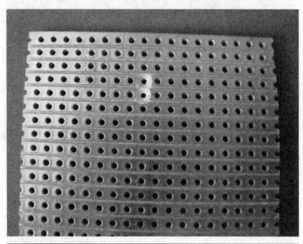

Figure 1-3 Stripboard tracks that have been cut

Chapter 1 ■ Read This Before You Start Any Projects

Please Read This Before Using Your Soldering Iron

Before you get started, here are some important safety points to remember when using a soldering iron to build your projects:

- Make sure that the room you are using to build the projects is well ventilated and properly illuminated.
- Avoid breathing in the toxic solder fumes when soldering. I recommend that you purchase a solder fume extractor that is designed to remove the fumes.
- Soldering irons become very hot. If you touch the metal tip or the shaft, you will burn yourself.
- You may want to investigate the use of soldering gloves.
- Molten solder is also very hot and can burn you.
- Soldering irons can be a fire hazard if they are left unattended, or if they are left on or nearby flammable materials.
- A hot soldering iron can also melt both flammable and nonflammable materials so always place a hot soldering iron in your soldering iron stand when it is not in use (there is a soldering iron stand shown in Figure 1-15 and you can purchase these from one of the suppliers mentioned in the appendix of this book).
- Use lead-free solder to build your projects. I use a lead-free solder which is supplied on a reel and has a 1/32" (0.7mm) diameter. Modern solders avoid the use of lead and are made up of a mixture of silver, copper, and tin; the type I used is made up of around 95.5% tin, 3.8% silver, and 0.7% copper.
- Always wear proper protective goggles or safety glasses to protect your eyes from stray solder splashes when building your projects.
- Always switch your soldering iron off and wash your hands after you have finished building your project.

1. If you start with a large piece of stripboard, you will probably prefer to cut the stripboard to the correct size using a small hacksaw (see Figure 1-4) and then use a fine-grade sandpaper to remove any rough edges on the stripboard. Ideally, you should purchase a piece of stripboard that is the correct size for the project so that no (or hardly any) cutting is required, especially if you don't have a hacksaw and vise available.

2. If you need to cut copper tracks, use a small drill bit fitted into a wooden handle (or use a spot face cutter) to make cuts manually in the tracks where required, as shown in Figure 1-5. Don't put too much pressure on the stripboard

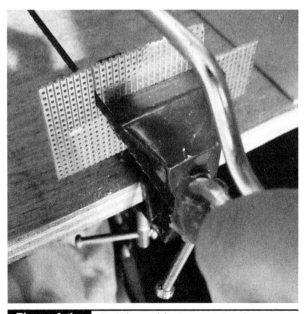

Figure 1-4 Stripboard being cut to size

Figure 1-5 Tracks being cut

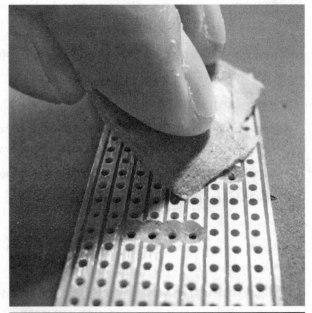

Figure 1-6 Stripboard being cleaned

when twisting the drill bit or else you will create a hole in the board or may even break it.

3. Gently sand the copper side of the stripboard using a fine-grade sandpaper to smooth the cut tracks and to clean the board in preparation for soldering, as shown in Figure 1-6 (also see the "Soldering Tips and Techniques" section later in this chapter). There are also special polishing blocks available that you may prefer to use to clean the copper side of the stripboard. Be sure that you don't sand the copper side too vigorously; otherwise, you will damage the copper surface, possibly breaking the tracks. Remove any dust and debris from the stripboard before you begin to solder; do this by carefully tapping the edge of the board on your bench and then wiping the board with a dry rag.

4. If the project uses integrated circuits (ICs), then I recommend that you use dual in-line (DIL) IC sockets instead of soldering the ICs directly into the stripboard. That way, you can remove and replace faulty ICs without having to desolder them from the stripboard. Fit and solder any IC sockets first, as shown in Figure 1-7.

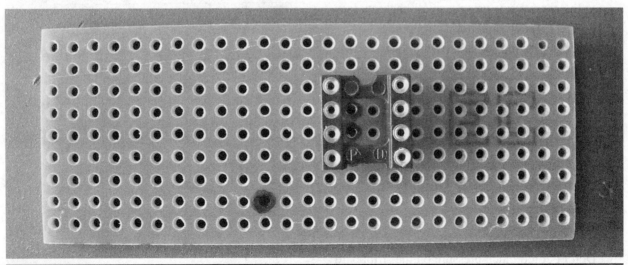

Figure 1-7 Solder the IC sockets first.

TIP Use a nonflammable insulation tape to help hold the components and wire links in place on the noncopper side of the stripboard while you are soldering.

5. Fit and solder the wire links, as shown in Figure 1-8. You need to use solid tinned copper wire for the links. The wire that I used for the projects in this book was RS Components part number 355-079; you need to ensure that the wire that you use has a diameter that is small enough to fit through the holes in the stripboard (while making sure that it has a current rating of at least 1 amp). You can bend the wire links slightly to prevent them from falling out before you start to solder them in place. Once you have soldered the links, you can cut the excess wire to size by using small wire cutters. Make sure that you don't cut into the solder joint.

CAUTION When cutting wire links and component leads to size, make sure that you hold the wire as you cut it. Otherwise, the excess wire could fly through the air, which can be dangerous.

6. Fit the resistors, capacitors, transistors, and any other components (see Figure 1-9). Again, remove excess leads in the same way as you

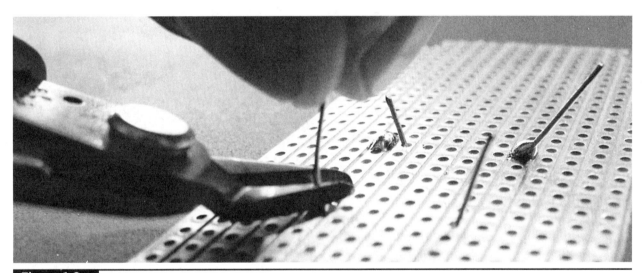

Figure 1-8 Cutting the wire links to size

Figure 1-9 Fully populated stripboard

removed the wire links, by holding and cutting them to size with your wire cutters. Finally, you can solder any battery connections or flying leads to the stripboard.

7. Once all the components are fitted, make sure that there is no solder bridging any of the gaps in between adjoining tracks (see Figure 1-10). If this occurs you will need to remove the solder by following the guidelines outlined later in this chapter.

8. Check to make sure that your stripboard layout matches the layout diagram for the project that you are working on (see Figure 1-11). If it does not match, then you may need to modify the stripboard accordingly.

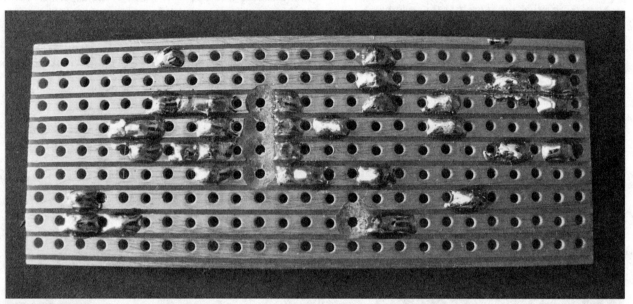

Figure 1-10 Check the solder joints.

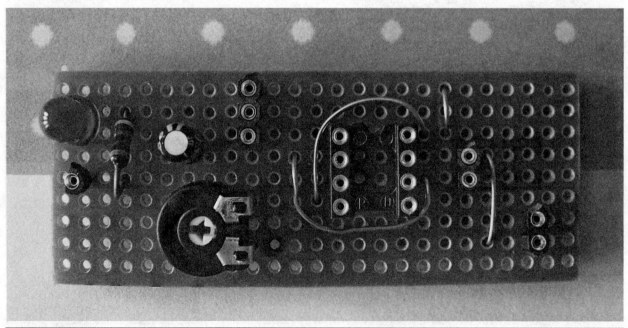

Figure 1-11 The circuit is now complete and needs to be checked.

9. Once you are happy with the stripboard layout and the quality of the solder joints, insert the ICs into their sockets, making sure that they are inserted in the correct orientation. The dot on an IC helps you to identify pin 1. If a dot is not present, then the semicircle on an IC denotes the top of the IC; pin 1 can then be identified because it is the top-left hand pin of the IC (see Figure 1-12). You are now ready to power up the circuit to see if it works by following the detailed instructions outlined in each project chapter.

10. If the circuit works right away, then you have successfully built the circuit and are finished with the task. If the circuit doesn't operate as expected, then continue with the following fault-finding steps.

11. Remove the battery immediately and visually check the board to make sure that it is constructed in accordance with the layout diagrams, and that there are no dry solder joints or solder splashes across adjoining tracks.

12. After performing any remediation in step 11, if the circuit still doesn't operate, use a multimeter to check the voltage levels at various points on the board to ensure they are as expected, as shown in Figure 1-13. It is sometimes best to remove the IC from its socket while checking the board over. You can also check the resistance of the resistors, although it is unlikely (although not impossible) that the resistors are faulty. The problem could be a faulty, damaged, or incorrectly inserted IC, electrolytic capacitor, LED, or transistor. If the circuit uses a PIC microcontroller, check that it has been programmed with the correct hexadecimal ("hex") code. Unfortunately, fault-finding is not an exact science, and sometimes it is a matter of trial and error. Understanding how the circuit operates does make the process a lot easier.

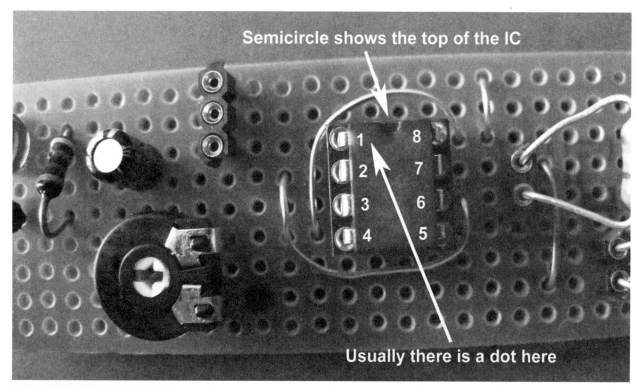

Figure 1-12 Make sure the ICs are correctly oriented

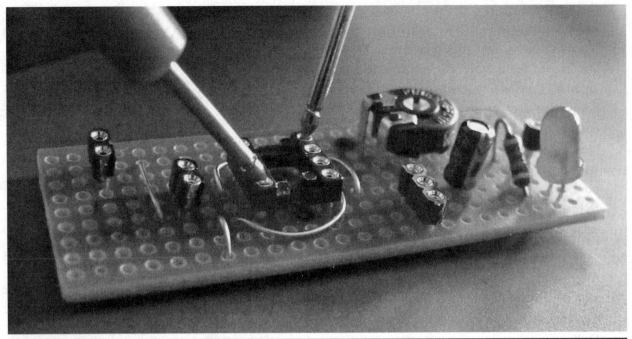

Figure 1-13 Fault-finding the circuit

TIP Fault-finding can be one of the most frustrating aspects of working with electronics, but it can also be very satisfying when you find and fix the problem, so just persevere and you will eventually succeed at creating a working project.

Soldering Tips and Techniques

This section provides some tips on choosing a suitable soldering iron and associated equipment. It assumes that you do not already own any soldering equipment; however, even if you do it is still worth a read because it then goes on to discuss some useful soldering techniques.

Choosing Soldering Equipment

If you don't own a soldering iron but want to purchase one, your first consideration should be which type you need. Soldering irons come in various shapes and sizes, and some are powered by AC electricity while others are portable and are powered by battery or gas. When building the types of projects presented in this book, I use a soldering iron with a small to medium size solder tip. The electronics suppliers listed in the appendix offer a good selection of soldering irons and accessories for hobbyists. You also need to make sure that the operating temperature of your soldering iron is going to be hot enough to melt the type of solder that you decide to use to build your projects. I used an electric-powered 18W soldering iron to build the projects in this book.

You should also use a soldering iron stand, which will help you to avoid placing the hot iron on your work surface. Soldering iron stands are usually supplied with a sponge; you can dampen the sponge before you start building the project, and then wipe the soldering iron tip across the damp sponge whenever you need to remove excess solder from the tip.

The type of solder that you use for your projects is also important. Solder is normally supplied on reels and contains a flux that runs through the center of the solder. The flux helps to clean the copper surface of the board that you are soldering and aids in creating a better solder joint. To build the projects in this book I used a lead-free solder, which again can be purchased from reputable electronics suppliers; the type that I used is supplied on a reel and has a 1/32" (0.7mm) diameter, and it is made up of 95.5% tin, 3.8% silver, and 0.7% copper.

It is also worth investing in a desoldering tool (sometimes known as a solder sucker) and also some desoldering braid. These items help you to remove any unwanted solder from the board if you make any constructional errors. These two items are discussed in more detail shortly.

Practicing Your Soldering Techniques

Now it is time to practice your soldering techniques. Like anything else, practice makes perfect, so before you build the projects in this book, try several practice runs using some scrap pieces of stripboard and some old components or tinned copper wire. The following steps should help you to create a good, clean solder joint every time.

It is worth turning on your soldering iron before starting to build your project and placing it in your soldering iron stand; the iron can then be warming up while you prepare the stripboard.

1. Prepare the stripboard before soldering by ensuring that the copper surface is clean and shiny, as shown in Figure 1-14. The copper surface can become dull over time, and grease off your fingers can collect on the copper surface, both of which can help to cause dry solder joints. You can use a polishing block or fine-grade sandpaper to clean the board. Sanding the copper side of the board too

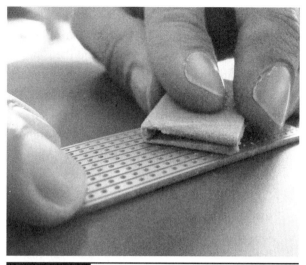

Figure 1-14 Cleaning the board

vigorously will damage the copper surface and could damage the tracks, so be gentle. If the stripboard is new and shiny, then you may not need to clean it at all. Once the surface is clean and dust free, it is time to insert a component onto the clean stripboard.

2. Wipe the hot soldering iron tip briefly on the damp sponge to remove any solder blobs (see Figure 1-15).

Figure 1-15 Cleaning the tip

3. After the soldering iron reaches its operating temperature, use the tip of the soldering iron to heat up the component lead and the stripboard for a few seconds, making sure that the soldering iron is angled at around 45 degrees to the stripboard (see Figure 1-16). How you determine the operating temperature of your soldering iron depends on the type of iron that you are using. Basic soldering irons do not normally have an indicator light or display showing the temperature like some of the more expensive models. If you are using a basic soldering iron (like I do) then you just need to wait for a few minutes after switching the soldering iron on, and then touch the tip of the iron on a piece of solder to see if it melts. If the solder doesn't melt quickly then the soldering iron is not hot enough.

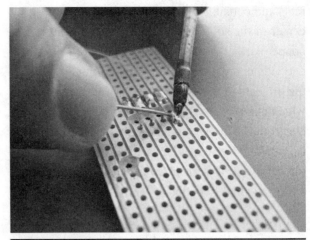

Figure 1-17 Feeding the solder onto the component lead

5. Once the solder joint is created, remove the soldering iron and solder to allow the molten solder and component lead to cool down. Don't move the board or the component while the joint is cooling, as this can weaken the joint. You should end up with a neat joint similar to the one shown in Figure 1-18.

Figure 1-16 Angling the tip

Figure 1-18 A neat solder joint

4. Use your free hand to "feed" the solder onto the junction that is created by the hot tip, the component lead, and the stripboard, which will cause the solder to melt onto the component lead and the copper strip, as shown in Figure 1-17.

CAUTION Too much heat can damage electronic components, so leave the solder tip on the lead long enough to create a neat joint, normally a maximum of a few seconds, but not long enough to damage the component.

The solder joint should be clean and shiny, not dull and misshapen like the joint shown in Figure 1-19. Poorly soldered joints and dry joints are physically weak and don't conduct very well, which will make the final circuit unreliable and can cause the circuit not to operate at all. If you have followed the instructions so far you should have created a neat solder joint; if your joint looks like Figure 1-19; however, this could be because your soldering iron is not hot enough or because the copper tracks have not been cleaned properly. Sometimes a component lead may be dirty, which will cause the same problem; this can be resolved by gently sanding the lead with sandpaper.

Figure 1-20 Too much solder can bridge the tracks.

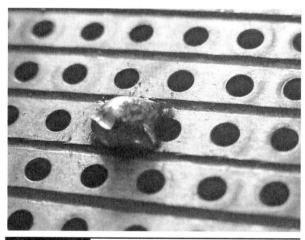

Figure 1-19 A poor solder joint

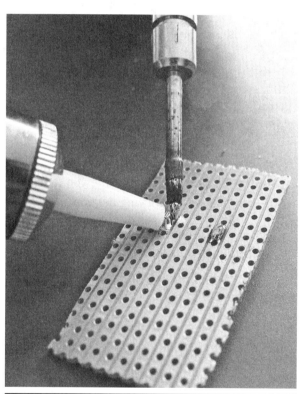

Figure 1-21 Using a desoldering tool to remove the solder

6. If you apply too much solder to the joint, stray solder might bridge the stripboard tracks, as shown in Figure 1-20. You need to resolve this before you do anything else, in either of the following ways:

    - Use a desoldering tool to remove the stray solder. Heat up the solder joint with the soldering iron and use the desoldering tool to suck up the hot solder, as shown in Figure 1-21.

    - Put some desoldering braid on top of the joint, heat up the braid and the joint with the soldering iron for a few seconds, as shown in Figure 1-22, and then remove the iron and the braid. The braid should soak

Figure 1-22 Using desoldering braid

up the stray solder, enabling you to remove the solder by removing the braid.

7. Once you have a clean joint, repeat this soldering process for the rest of the joints on the stripboard.

Antistatic Precautions

Some sensitive electronic components can be damaged by static electricity, especially CMOS integrated circuits. As you are probably aware, high levels of static can build up in your body just by walking across a carpet and can also be generated by your clothing. If you have ever received an electric shock from touching someone, you realize how much of a charge can be easily built up in the human body. This electric charge can be transferred to electronic ICs and can damage them.

You should remove any static electricity from your body before you start to build your projects, to avoid damaging your components. The way to do this is to ground yourself; this allows any static electricity that has built up in your body to flow away into the ground. The proper method of protecting sensitive components from static is to build your projects on a proper antistatic mat and to wear the conductive wrist strap that connects to the mat. This type of mat has a connection that has to be permanently grounded; in the U.K. this is normally via a cable and a plug which grounds the mat via the earth pin of the household AC electric supply. This means that you are always permanently grounded while you are working, which ensures that static isn't able to collect in your body.

Working with Breadboard

If you decide that soldering is not for you, you don't need to close this book and look for another electronic design book. That would be a shame, because you would be missing out on some brilliant LED projects! There is no reason why you can't build the circuits in this book using breadboard, which is an alternative method of building circuits that doesn't require you to pick up a soldering iron. The circuits may not be as compact as they would be on stripboard, but they work just as well. Although this book does not show breadboard layouts for the complete projects, it does show some experimental circuits that use breadboard. A photograph of a typical piece of breadboard is shown in Figure 1-23. If you use breadboard there is no soldering required because the pins and leads of the electronic components simply push into the holes. They remain in place because the electrical contacts that sit inside the breadboard "grab" the leads and hold them in place, and you can simply remove the components from the board by carefully pulling them out.

Programming PIC Microcontrollers

While some of the early projects in this book use standard integrated circuits, most of the projects from Chapter 12 onwards use the PIC Microcontroller, which is manufactured by Microchip Technology, Inc. More details about how PIC microcontrollers work are provided in the corresponding projects, but basically this component is a programmable integrated circuit, which means you need to load a software code into the chip to make it work. The operation of the assembly code and hex code is explained in each project chapter in which the PIC Microcontroller is used.

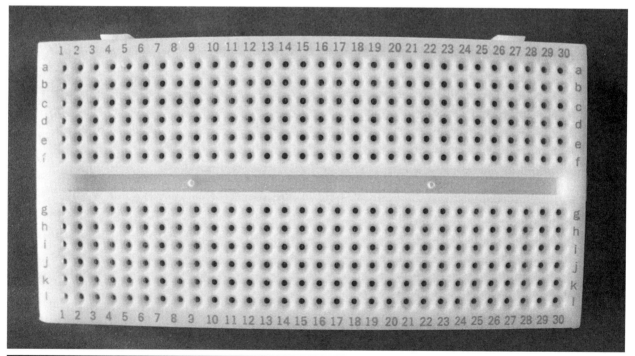

Figure 1-23 You can use breadboard to build your projects if you prefer not to solder.

NOTE You can download both the assembly code and the hex code software from the McGraw-Hill website at www.mhprofessional.com/computingdownload (click the title of this book).

To program the chip, you need a dedicated programmer solution. Various types are available. I used Microchip's PICkit 2 Development Programmer/Debugger and software to program the PIC Microcontroller projects in this book. The PC software to download the hex code into the microcontroller is provided with the programmer. There are six output pins from the PICkit 2 programmer, configured in the following order (pin 1 of the programmer is denoted by the white triangle):

- Pin 1: $\overline{\text{MCLR}}$
- Pin 2: Vdd Target (+V)
- Pin 3: Ground (0V)
- Pin 4: Data
- Pin 5: Clock
- Pin 6: Aux

For the purpose of programming the PIC Microcontroller, you need to use pins 1, 2, 3, 4, and 5.

You need to be able to connect the programmer to the PIC Microcontroller so that you can program it. To help you out, I decided to build a homemade interface on a piece of breadboard. Figure 1-24 shows the breadboard layout that I built to program the microcontroller projects in this book. Using this layout, you connect the PICkit 2 programmer to your PC by using a USB cable, and then you plug the programmer into the single in-line (SIL) pin header fitted to the breadboard. After you

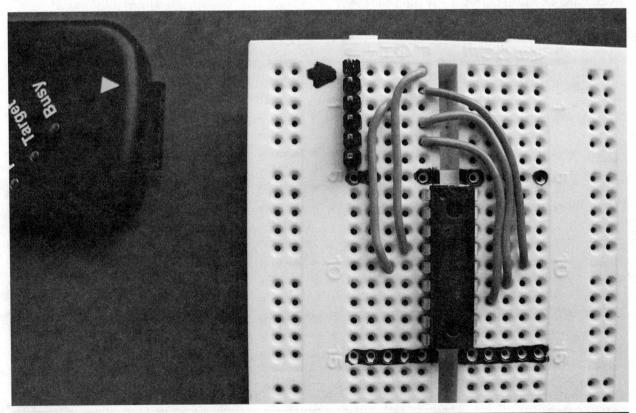

Figure 1-24 Homemade PIC Microcontroller programmer breadboard. Note: Make sure that the holes in your breadboard are large enough to accept the pin header. *Do not* force the pins into the board; otherwise, you could damage the board and short out the pins. Alternatively, you may consider building a programming interface using stripboard instead.

connect the programmer to the breadboard, you can program the hex code into the PIC Microcontroller via the software. More details about how to program these devices are provided in Chapter 12.

Resistor Color Codes

Table 1-1 outlines the most common color codes for resistors with four colors (E12 type resistors). The first three colors of the code allow you to calculate the resistor value in ohms (Ω) and the fourth color is the % tolerance of the resistor. Table 1-1 and the tolerance list help you to quickly calculate resistor values.

The fourth color denotes the tolerance of the resistor:

- Brown: 1%
- Red: 2%
- Gold: 5%
- Silver: 10%

So, for example, by looking at Table 1-1, you can see that a resistor that has a color code of Yellow, Violet, Brown, Gold has a resistance of 470Ω (ohms), and the preceding list indicates that it has a 5% tolerance. If you are ever unsure of the resistance value of a resistor, use your multimeter to check its value.

Circuit Diagrams and Stripboard Layouts

Each chapter contains a circuit diagram that shows you the layout of each circuit. This book assumes that you are familiar with schematic diagrams, but it also provides a detailed description of the circuit operation, so if you aren't familiar with schematic diagrams, you should be able to adapt as you go by referencing the diagrams as you read the text. Each

TABLE 1-1 Four-Color Resistor Value Reference Chart

| | | | | Third Color (Multiplier) | | | | | | |
| | | | | x 0.1 | x 1 | x 10 | x 100 | x 1000 | x 10,000 | x 100,000 | x 1,000,000 |
First Color		Second Color		Gold	Black	Brown	Red	Orange	Yellow	Green	Blue
Brown	1	Black	0	1Ω	10Ω	100Ω	1KΩ	10KΩ	100KΩ	1MΩ	10MΩ
Brown	1	Red	2	1.2Ω	12Ω	120Ω	1.2KΩ	12KΩ	120KΩ	1.2MΩ	
Brown	1	Green	5	1.5Ω	15Ω	150Ω	1.5KΩ	15KΩ	150KΩ	1.5MΩ	
Brown	1	Gray	8	1.8Ω	18Ω	180Ω	1.8KΩ	18KΩ	180KΩ	1.8MΩ	
Red	2	Red	2	2.2Ω	22Ω	220Ω	2.2KΩ	22KΩ	220KΩ	2.2MΩ	
Red	2	Violet	7	2.7Ω	27Ω	270Ω	2.7KΩ	27KΩ	270KΩ	2.7MΩ	
Orange	3	Orange	3	3.3Ω	33Ω	330Ω	3.3KΩ	33KΩ	330KΩ	3.3MΩ	
Orange	3	White	9	3.9Ω	39Ω	390Ω	3.9KΩ	39KΩ	390KΩ	3.9MΩ	
Yellow	4	Violet	7	4.7Ω	47Ω	470Ω	4.7KΩ	47KΩ	470KΩ	4.7MΩ	
Green	5	Blue	6	5.6Ω	56Ω	560Ω	5.6KΩ	56KΩ	560KΩ	5.6MΩ	
Blue	6	Gray	8	6.8Ω	68Ω	680Ω	6.8KΩ	68KΩ	680KΩ	6.8MΩ	
Gray	8	Red	2	8.2Ω	82Ω	820Ω	8.2KΩ	82KΩ	820KΩ	8.2MΩ	

project also includes a stripboard layout diagram and close-up photographs, which will help you to build each of the circuits. For your reference, Figure 1-25 shows most of the common component symbols that are used in the schematic diagrams that appear in this book.

Now you are ready to learn about LEDs and to go through the process of building your first project. I hope that you enjoy the journey ahead.

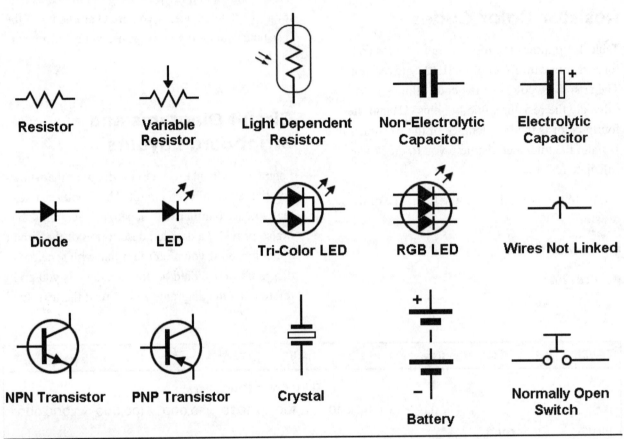

Figure 1-25 Schematic symbols

Warnings Before You Get Started

- Only build these projects if you feel confident in doing so. It is also recommended that you read through each project's description before you embark on building it; this will help you to understand how the circuit works and to plan the project out before you start.
- Many of the projects outlined in this book utilize bright and colorful flashing light effects. If you suffer from epilepsy or are affected by flashing lights, then please do not attempt to build these projects.
- The projects outlined are for experimental use only. The stripboard layouts have not been tested for electromagnetic compatibility (EMC) and are not intended for commercial use.
- Electronic components are small and can be a choking hazard, so always keep them out of reach of small children.

- Experimenting with electronics is fun, but don't be surprised if occasionally you accidentally connect components incorrectly and they become damaged. This is part of the experience and is sometimes to be expected, so be patient and carry on.

- In normal operation the components used in each project should not become hot. If you find that any of the components become hot after you power your project up, then remove the battery immediately and check the board for faults. Also check that you are using the correct component values and the correct battery voltage as outlined in the relevant project chapter.

Drilling and Cutting Guidelines

Some of the projects show you how to house the stripboard layouts into plastic enclosures and this may require you to perform some drilling or cutting. If you are not old enough to own an electric drill or you don't feel confident in using one, then you should either find an alternative method to house your electronic projects or ask a responsible adult to perform the drilling and cutting for you.

The following points are not exhaustive and should be the minimum precautions that you should take; each project also outlines further suggestions and guidelines:

- Always ensure that the part that you are drilling or cutting is held firmly in a vise.

- Always wear safety goggles or glasses to protect your eyes when drilling or cutting.

- Make sure that the drill bit is secured tightly in the drill before starting to drill the hole.

- Use a variable-speed drill, this will allow you to reduce the drilling speed to a controllable level—I use a battery-powered drill, which also means that there are no trailing power leads to watch out for when drilling.

- Drill small pilot holes in the plastic enclosure first—this can sometimes be done by hand without having to use an electric drill. You can do this by using a 1/8" (3mm) drill bit mounted in a wooden handle.

- Once a pilot hole is created, you can then create a bigger hole by using the correct size drill bit—again, sometimes this can be performed by hand depending on the thickness of the plastic enclosure. Be sure to mount the drill bit in a wooden handle if you do this.

- If you need to create large holes in the enclosure, use your drill to create the largest hole possible, and then carefully use a metal file by hand to widen the hole to the size required.

- Always keep your hands and fingers away from the part that you are cutting or drilling.

- Holes can be easily created in stripboard by hand, using a drill bit mounted in a handle as described above.

- Use a small hacksaw if you decide to cut into plastic or stripboard.

PART ONE
Illumination and Flasher Projects

CHAPTER 2

Basic LED Circuits: How to Make an LED Flashlight

TO CREATE BRILLIANT LED PROJECTS, YOU NEED A basic understanding of the components you are working with. Thus, this chapter first explains what light-emitting diodes (LEDs) are and how you can start to use them in your projects. This then leads to your first project, which shows you how to make an LED flashlight.

What Are Light-Emitting Diodes?

LEDs are relatively low-cost, solid-state illumination devices that have many advantages over their standard incandescent light bulb counterparts. These advantages include fast switching times, no filament to blow, low operating current, long operating life, and fairly cool operating temperatures. These key benefits make LEDs a popular choice for electronic designers, which is why LEDs now appear all around us in diverse products such as traffic lights, remote controls, TV screens, mobile phones, lighting solutions, and automotive applications, to name a few. LEDs have been used in commercially available electronic products for several decades and are now a permanent part of our everyday lives. More recently, "super bright" white LEDs are becoming a viable alternative to standard lighting in the home and on commercial premises, and with their low operating currents, they are being considered as a possible solution to reducing our electricity consumption in the future.

The range of LED components available to hobbyists and experimenters is quite varied and includes single-color (in various colors), bicolor, tricolor, ultra-bright, seven-segment numerical, infrared, ultraviolet, bar-graph, and dot-matrix displays. These LED devices are also available in various shapes and sizes, a selection of which are shown in Figure 2-1. You will encounter many of these components throughout this book.

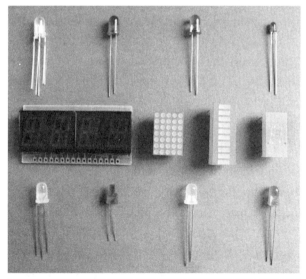

Figure 2-1   A selection of various LED devices

A standard *diode* is a semiconductor component that allows current to flow in one direction but not the other (we will explore diodes in later projects).

LEDs have similar properties to diodes but they also produce light when current flows through them.

The circuit symbol for a standard diode is shown here:

And this is the circuit symbol for an LED:

The double arrows indicate that the diode illuminates.

How to Illuminate an LED

The first thing you need to understand about LEDs before you build your first project is how to illuminate an LED. It is not the same as illuminating a standard incandescent light bulb, which will illuminate when connected directly across a suitable voltage source, regardless of which way round the two leads of the bulb are wired across the voltage source. If you use this method with an LED, it may not illuminate at all and you could destroy it.

Figure 2-2 shows a typical LED. As you can see, the LED has two connections, similar to a incandescent light bulb, but unlike an incandescent light bulb, the LED has a positive lead and a negative lead, which need to be wired in the correct way across a voltage source; otherwise, as previously mentioned, the LED will not illuminate and may become damaged. These two connections are called the *anode* (+) connection, which is sometimes denoted with an *A* on a circuit diagram, and the *cathode* (−) connection, which is sometimes denoted by a *K* on a circuit diagram.

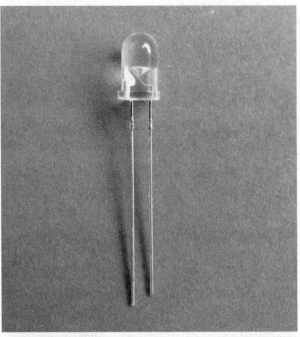

Figure 2-2 The cathode (−) is normally next to the flat side of the LED. Refer to the caution at the end of this chapter.

The LED functions like a normal diode, allowing current to flow in only one direction. When the diode is connected properly so that it allows current to flow through it—that is, the anode leg of the LED to the positive lead of the voltage source and the cathode leg of the LED to the negative lead of the voltage source—it is said to be in *forward bias*. If the LED is connected the wrong way around, this is known as *reverse bias*. If the forward or reverse bias voltages exceed the maximum voltages specified on the LED datasheet, the LED will be destroyed and irreparably damaged. Applying too high a voltage to an LED in either forward or reverse bias can be quite destructive and is not recommended. Without going into the science of how an LED works, you just need to know that illumination is produced when the LED is in forward bias.

Normally, you can identify which lead is which simply by looking at the LED. Looking again at the LED in Figure 2-2, the longer lead of the LED is the anode (+), which is connected to the smaller electrode inside the LED. The shorter lead is the cathode (−), which is connected to the larger electrode inside the LED. The cathode is also usually positioned next to the flat edge of the LED.

The other thing to remember is that if you connect the LED straight across the battery, you will damage the LED. Some LEDs, such as flashing LEDs, are designed to be connected directly across a voltage source, but the correct method of powering up a standard LED is to ensure that you wire a resistor in series with the device, to limit the current flowing through it. Have a look at the basic LED circuit diagram in Figure 2-3.

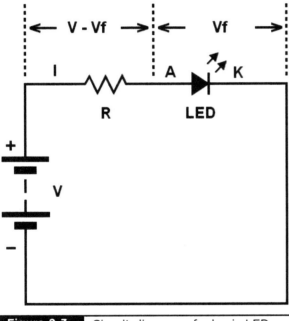

Figure 2-3 Circuit diagram of a basic LED circuit

There are three main considerations when including an LED in a circuit:

- The resistance (R) value in ohms (Ω) required for the series resistor.
- The wattage (W) rating of the series resistor.
- The total drive current available from the circuit controlling the LED(s).

The total drive current is also known as the source current and is covered in more detail in later chapters. The resistance and wattage calculations for the series resistor are covered in the following two sections.

Series Resistor Calculations

The resistance (R) value of the resistor in Ω is calculated using the following formula:

$$R = (V - V_F) / I_F$$

where V is the supply voltage, V_F is the typical forward voltage drop of the LED, and I_F is the current allowed to flow through the LED. Both V_F and I_F are specified on the LED datasheet.

NOTE The mA value of I_F needs to be converted to amps in all of these equations.

LED datasheets normally show typical and maximum I_F values, but it is not necessary to use the maximum value when calculating the resistor value. In fact, you should get a good level of brightness from most LEDs with an I_F of 10mA (milliamps). The V_F and I_F values will vary depending on the shape, size, and type of the LED. So, for example, a standard 5mm red LED might have a typical V_F rating of 2.8V and a typical I_F rating of 20mA. If the supply voltage of the circuit is 5V and you want to limit the current to 15mA, then the value of the series resistor would be calculated as 146.7Ω by using the following formula:

$$R = 146.7Ω = (5V - 2.8V) / 0.015A$$

CAUTION If your LEDs are being powered by an integrated circuit (IC), then the I_F value that you use in the series resistor calculations must also be *lower* than the maximum source current that is available from the IC. This is especially important if you need to recalculate the LED series resistor values for the projects in this book. For example, if an IC output pin can only source a maximum of 4mA to power an LED, then the value that you use for I_F in the formula should be lower than 0.004A. Source currents are discussed in more detail in each chapter of this book.

Referring to Table 1-1 in Chapter 1, you can see that there isn't a standard four-band resistor with this value, so you should select a *higher*-value resistor that is available to you, in which case you need to consider the tolerance and the wattage of the resistor used. For example, in this case, if you were to use a 150Ω resistor with a ±5% tolerance, the 150Ω resistor could actually have a value of 142.5Ω, which is lower than the 146.6Ω calculated originally. In practice, this may or may not matter too much. If the resistor value is 142.5Ω, then the current draw through the LED would be 15.4mA, which is not much higher than the 15mA that you wanted to achieve. If the LED can support this (in the example, the I_F rating is 20mA), then you are okay. If you were to use a 150Ω resistor with a tolerance of 2% instead, then the minimum tolerance value of the resistor would change to 147Ω, which is almost the same as the value you actually require. The purpose of this example is to highlight the fact that there are a number of factors that you need to consider when deciding on component values.

The nice thing about electronics is that the values of components can be tweaked slightly to create similar results, so if you do not have a 150Ω resistor available but you have a 180Ω or 220Ω resistor in your toolbox, then either should work fine. So what happens if you reduce the LED current in the calculation to 10mA?

$$R = (5V - 2.8V) / 0.010A$$

By using the formula, you can see that the resistance value required then increases to 220Ω. This demonstrates that increasing the resistance of the series resistor reduces the current flowing through the LED, and this in turn reduces the brightness of the LED.

TIP The V_F rating of LEDs can vary between batches and can alter slightly depending on the amount of current flowing through it. This could mean that the actual current flow through an LED may be slightly different to the one you have calculated. You can check the exact current flow by building a circuit like Figure 2-3 on breadboard, using the type of LED, resistor, and voltage rating that you want to use in your project. The actual current flow can then be measured using a multimeter. You can then decide whether you need to alter your calculated resistor value slightly.

Series Wattage Calculations

Now that you know how to calculate the resistance, it's time to learn how to calculate the wattage of the resistor. The formula for calculating wattage is

$$Wattage = Volts \times Amps$$

Assuming a 5V supply is used and that the current drawn from a single LED is 15mA, the power dissipation of the whole circuit is

$$W = 5V \times 15mA = 0.075W$$

Based on an LED with a V_F of 2.8V, the actual power dissipated through the resistor can then be calculated as $(V - V_F) \times 15mA = (5V - 2.8V) \times 15mA = 0.033W$.

However, the easiest way to calculate the wattage value required for an LED series resistor is to use this formula:

$$W = I^2 \times R$$

where I = the amount of current flowing through the resistor and R = the resistor value.

Using the 15mA (I_F) example from earlier in the chapter, the calculation is as follows:

$$W = 0.015A \times 0.015A \times 150Ω = 0.033 \text{ watts}$$

Or worst case with a ±5% tolerance resistor (+5% being the worst case in this instance):

$$W = 0.015A \times 0.015A \times 157.5Ω = 0.035 \text{ watts}$$

So it is safe to use a resistor with a wattage rating of 0.5W in this example. If you do not have a 0.5W resistor, you can go higher and use a 1W-rated resistor instead.

Don't forget that if you increase the resistance value of the series resistor, this will affect the total current and wattage rating of the circuit.

Illuminating Multiple LEDs

There might be situations in which you want to illuminate more than one LED from a single battery. To ensure an even current distribution through the circuit and an even light output from the LEDs, it is recommended that you wire the LEDs in parallel with each other, and that you give each LED its own series resistor. The circuit diagram in Figure 2-4 shows how you could power four LEDs from one single power supply.

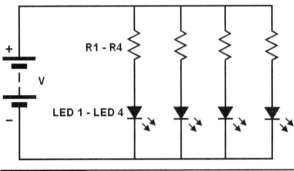

Figure 2-4 Circuit diagram of four LEDs wired in parallel with four resistors

You will see that each LED has its own resistor wired in series with it, so assuming that each LED draws 15mA from the supply, then the total current drawn from the supply is 60mA, but because the current flowing through each resistor is only 15mA, the wattage rating of each resistor can still be rated at 0.5W (using the example circuit calculation used earlier).

In theory, it is possible to eliminate three of the four resistors and connect the four LEDs together, as shown in the diagram in Figure 2-5.

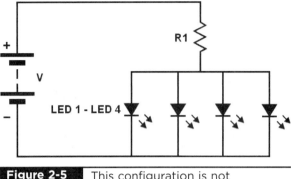

Figure 2-5 This configuration is not recommended.

However, if you were to use this method, the current flowing through the single resistor would increase (in this example to 60mA), which means that the wattage rating of the resistor would likely also need to increase. Another reason why this configuration is *not* recommended is that the V_F rating of the same type of LED may vary between batches, in which case the current flowing through each LED will vary, and this will also affect the amount of light output from each LED.

You do not need to be too concerned at this stage about the total drive (source) current available from the supply to power the LEDs, because the circuits shown so far are being powered directly from a battery and the current draw is fairly low. Most standard portable batteries should be able to supply 60mA, but you do need to think about battery consumption and how long the battery will last before running flat if the LEDs are left on continuously.

Now that you understand the basics of LEDs, you can start to build your first LED project.

Project 1
LED Flashlight

In your first project, you will use the concepts covered thus far in the book and the circuit previously described in this chapter to build a basic LED flashlight like the one shown in Figure 2-6.

Figure 2-6   LED flashlight

PROJECT SPECIFICATIONS

- The flashlight is handheld, fairly compact, and portable.
- The light source is two high-intensity white LEDs.
- A single push button activates the LEDs.
- The flashlight provides a good level of illumination in the dark.
- The power supply is 6 volts.

The circuit diagram for the LED flashlight is shown in Figure 2-7; it should look fairly familiar because it is a modified version of the basic LED circuit shown earlier in this chapter in Figure 2-4.

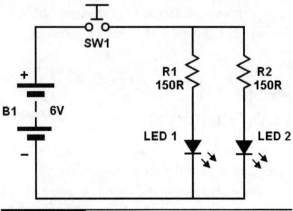

Figure 2-7   LED flashlight diagram

The power supply used in this circuit is 6V, which is provided by four 1.5V AA batteries wired in series.

The circuit is very straightforward: when the push button switch (SW1) is operated, the current is allowed to flow from the battery through the two resistors to illuminate the two LEDs. The LEDs chosen for this application are two 5mm high-intensity white LEDs. The light output of LEDs is measured in millicandelas (mcd), and this value varies depending on the type of LED. For example, a standard 5mm diffused LED might normally have a light intensity of less than 100mcd, whereas the high-intensity white LEDs used in this project have a light output of 6000mcd. This means that they provide a really good level of illumination, which is ideal for this application.

The V_F rating of this type of LED is also normally higher than that of standard color LEDs. A typical 5mm red LED may have a V_F value of 2.8V, whereas the LED used for this project has a typical V_F rating of 3.2V. This type of LED has an absolute maximum I_F rating of 30mA. I decided to calculate the series resistor based on a 20mA I_F, which is shown as a "typical" value on the datasheet. Using the resistor calculations outlined earlier, you can see that this means the resistor value is calculated as 140Ω, and therefore you can use a series resistor with a value of 150Ω. Even with a battery voltage of 6.6V (which may be the case with fresh batteries) and a 142.5Ω resistance (taking into consideration the resistor tolerance), the current through the LED will still only be around 24mA, which is below the 30mA maximum value allowed for this LED.

TIP If you use LEDs that have V_F and I_F values that differ from those of the LEDs that I have used to build the projects in this book, then you may need to alter the resistance and wattage values of the series resistors that you use in your projects. You can use the standard LED resistor formulas described in this chapter to calculate the resistance and wattage values required.

SW1 is a momentary push-to-make switch (aka normally open switch, or NO switch), which means that you need to keep your finger pressed on the button to maintain the contact and keep the flashlight activated. The flashlight is automatically switched off when you release the button, thus saving battery power. You need to make sure that the switch used in the circuit is capable of supporting the total current draw of the LEDs. This is especially important if you decide to modify the project to drive more than two LEDs.

Parts List

NOTE The Supplier and Part Number column of the following table lists specific parts that I used in this project. Refer to the appendix for additional details about acquiring your parts.

The parts you'll need for the LED flashlight project are listed in the table below.

How to Make the Flashlight

No stripboard layout is required for this project; the circuit is so straightforward that you can simply solder all the components together to build your flashlight. Simply follow these steps:

1. Drill three holes in the body of the enclosure (not in the lid), as shown in Figure 2-8. The two holes that are positioned together are for the LEDs, and the other single hole is for the switch. Take the usual safety precautions when drilling, including wearing safety glasses, and make sure that the enclosure is held firmly in a vise while you are drilling it.

PARTS LIST

Code	Quantity	Description	Supplier and Part Number
SW1	1	Single-pole normally open panel mount switch (100mA)	RS Components 133-6502
R1/R2	2	150Ω 0.5W ±5% tolerance carbon film resistor (this value may need to be altered, as described earlier in the chapter)	—
LED 1 LED 2	2	5mm high-intensity white LED (6000mcd) V_F (typical) = 3.2V, I_F (max) = 30mA	RS Components 668-6338 (pack of 10)
B1	1	AAA battery holder (four AAA batteries)	RS Components 512-3568
Hardware	4	AAA battery (1.5V)	—
Hardware	2	LED clips (to suit 5mm LEDs)	—
Hardware	1	Small, narrow enclosure, approximately 4.88" (124mm) long x 1.3" (33mm) wide x 1.18" (30mm) deep	Maplin FT31
Hardware	—	Self-adhesive sponge tape, one cable tie approx. 4" (101mm) long x 0.1" (2.5mm) wide, and one cable tie base	—

Figure 2-8 Drilled enclosure for the LED flashlight

The hole for the switch needs to be large enough to accept the switch that you decide to use, and the two LED holes need to be large enough to accept the LED clips.

2. Cut two lengths of self-adhesive sponge pads and insert them into the box where the battery holder will sit, to stop the holder from moving around in the box. Then insert the battery holder in position and fit an adhesive pad above the holder. Secure the battery holder cables to the cable tie pad using a cable tie, while leaving enough slack in the cable to enable you to remove the holder and insert the batteries when you are ready (do not fit the batteries yet). See Figure 2-9.

3. Solder a resistor to each of the anode legs (+) of the LEDs (the anode is normally the longer lead). To do so, tin each LED and resistor leg with a coating of solder first and then apply heat to the legs with the soldering iron to join them together, as shown in Figure 2-10. Use pliers (rather than your fingers) to hold the resistor legs to the LED legs when soldering because the leads become very hot.

4. Insert the LED clips into the end of the enclosure and carefully insert the LEDs into the clips. This is normally done by pushing the head of the LED into the clip until it clicks into place, but the method may vary

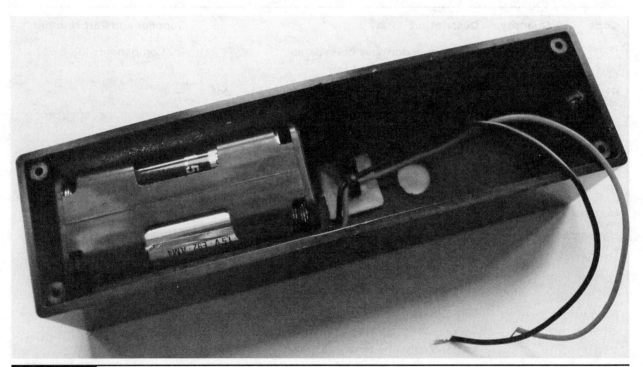

Figure 2-9 Securing the battery holder

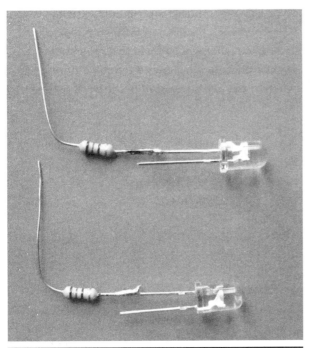

Figure 2-10 Solder the resistors to the LEDs.

depending on the type of clip used. Insert the switch, and then solder all the components together, as shown in Figure 2-11. Solder the positive (red) battery cable to the switch, and then twist together the two unconnected legs of the two resistors and solder them to the other leg of the switch. Finally, join together the two cathode (−) legs of the LEDs and solder them to the negative (black) battery cable.

CAUTION Make sure that you do not touch the sides of the plastic enclosure with the hot soldering iron when connecting and soldering the wires together.

5. Once you are happy that everything is wired up correctly, fit the four AAA batteries into the battery holder.

6. Press the push button switch (SW1) to illuminate the two LEDs. Do not look directly at the LEDs, as the light output is very bright. If the LEDs do not illuminate or are only dimly lit, then remove the battery and recheck the circuit before you proceed any further.

7. If the circuit works as expected, fit and screw down the lid of the enclosure. The finished unit should look similar to the one shown in Figure 2-12.

Figure 2-12 The finished LED flashlight

It is now time to try out your homemade flashlight in the dark. You should see that the two ultra-bright LEDs provide quite a good level of illumination for such a small flashlight. The total current draw is also fairly low, at around 40mA, and with intermittent use, the batteries should

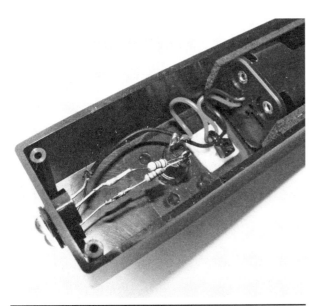

Figure 2-11 Solder the components together.

provide many hours of service. One thing to note is that the enclosure that I used is not waterproof, which means that unfortunately you cannot use the flashlight in the rain.

Congratulations, you have built your first LED project!

CAUTION The position of the anode and cathode leads of some LEDs doesn't always follow the convention outlined earlier in this chapter. It is therefore recommended that you double-check the datasheet of the LED that you use in your projects to identify which lead is which.

NOTE I used a push button switch which I salvaged from an old piece of electronic equipment to build this project; this is why the photograph shows a slightly different switch to the one suggested in the parts list.

CHAPTER 3

An Alternative Way to Power an LED: "Green" Pocket LED Flashlight

IF YOU COMPLETED THE PROJECT IN THE PREVIOUS chapter, you've illuminated an LED and created a flashlight, which is a good start to your journey to understanding the many things that you can do with LEDs. In this chapter's experimental project, you'll utilize the basics of the LED flashlight from project 1 and add another element to it.

One of the benefits of using LEDs is their low current consumption relative to that of an incandescent light bulb. Although some bulbs provide a better light output than ultra-bright LEDs, the dual LED flashlight that you built in Chapter 2 provides sufficient light output for use as a flashlight. The flashlight that you built has a current consumption of only around 40mA using two LEDs. This means that the battery should last longer under continuous use compared to the same circuit using a small single 5V bulb, which typically has a current consumption of at least 60mA.

While thinking about the low current consumption of LEDs, I began to think about the potential of powering an LED *without* the use of a battery, thereby creating an even more environmentally friendly ("green") version of the LED flashlight. What if we could store some electrical energy in a flashlight to provide enough power to a single LED to give a decent light output for a few minutes of use at a time? The solution is to use a capacitor, as you'll see demonstrated in this chapter's project. First, though, an overview of how capacitors function is in order.

Capacitors

The common capacitor stores an electrical charge. A capacitor is constructed internally with two metal plates, each of which is connected to a component leg. These plates are insulated from each other using a material that does not conduct electricity. Figure 3-1 shows some typical capacitors that you may come across.

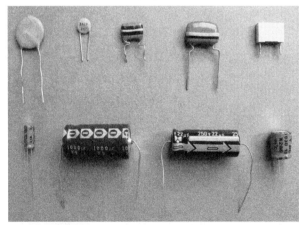

Figure 3-1 Photograph of various capacitors

31

Various types of capacitor are available, but the type we are interested in at the moment is the electrolytic type, examples of which are shown in the bottom row of Figure 3-1. This type of capacitor is *polarized*, which means that it has to be connected in the correct way around in the circuit. It has a positive connection and a negative connection, and normally has markings on it that denote which lead is which.

Capacitors also have a voltage rating, and you need to ensure that the capacitor's voltage rating is higher than the supply voltage of the circuit. A capacitor is normally used in electronic circuits to smooth a DC supply or to charge or discharge a voltage. The measurement of capacitance is farads (F), and capacitor values are measured in fractions of a farad, starting with the smallest value of picofarads (pF), moving up to nanofarads (nF), and finally microfarads (μF). The higher the value of the capacitor, the greater the electrical charge that can be stored.

CAUTION Although you are using only low voltages in your circuits, keep in mind that capacitors do retain electrical charge even when they are not connected to an electronic circuit. Ensure that you do not short out a capacitor (with your skin or any other implement) once it is charged, because the capacitor will dump its complete charge in one go, and this could cause a spark or an electrical shock. It is also important to ensure that the voltage rating of a capacitor is higher than the voltage of the circuit and that a polarized capacitor is connected in the circuit the correct way around; otherwise, it could leak or even explode.

Using a Capacitor to Power an LED

To give you an example of how an electrolytic capacitor works, this section shows you how to build some test circuits on a piece of breadboard. The components needed for the breadboard circuits are listed and described in Table 3-1.

TABLE 3-1	Breadboard Test Circuit Component List	
Quantity	Description	Supplier and Part Number
1	Breadboard	Maplin AD-100
2	1000ΩF 10V electrolytic capacitor	RS Components 684-1882 (pack of 5)
1	5mm red LED V_F (typical) = 2.0V, I_F (max) = 30mA	RS Components 228-5972 (pack of 5)
1	1.8KΩ 0.5W ±5% tolerance carbon film resistor	—
1	1F 5.5V electrolytic capacitor	RS Components 339-6843
1	AA battery holder (three AA batteries)	Maplin YR61R
3	AA battery (1.5V)	—
1	PP3 battery clip and lead	RS Components 489-021 (pack of 5)
—	Insulated jumper wires	—

First, build the simple circuit shown in Figure 3-2 on the piece of breadboard using a 1000µF 10V-rated capacitor, a 1.8KΩ 0.5W resistor, a 5mm red LED, and some interconnecting wires.

NOTE The breadboard that I used has internal connections which run horizontally across the board as you look at Figure 3-2. In total there are 60 individual rows, each of which contains 6 pin connections that are linked together. So for example in row 30 at the bottom of the board, pins 30g, 30h, 30i, 30j, 30k, and 30l are all connected together.

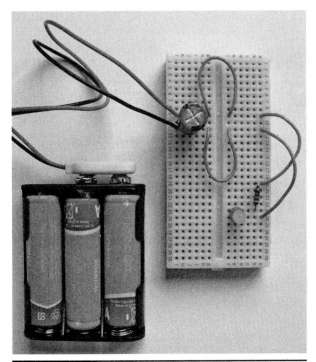

Figure 3-2 Breadboard capacitor circuit

Make sure that the capacitor is positioned in the correct orientation and then apply the 4.5V power source (provided by three AA batteries) for around 10 seconds, making sure that the + and – connections of the capacitor are connected to the matching polarities of the battery. The LED should illuminate as soon as the power is applied. After around 10 seconds, remove the power supply to the board and watch what happens. The LED should stay illuminated for a few seconds and then fade out. Basically, the battery started to charge the capacitor as soon as it was connected. It does not take long for the capacitor to charge, and at the same time the LED is receiving power and thus is illuminated. When the power is removed, the capacitor releases its charge through the resistor and illuminates the LED for a brief period until it is fully discharged.

Now add another 1000µF capacitor in *parallel* with the 1000µF capacitor, as shown in Figure 3-3, and then apply the 4.5v power source again for about 10 seconds and see what happens. This time you will see that the LED remains illuminated for a slightly longer period of time after you remove the power supply.

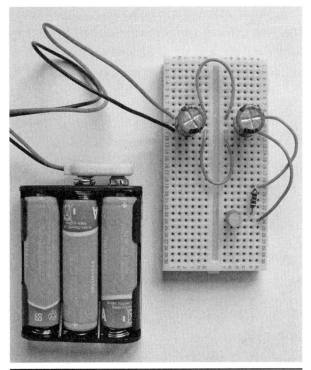

Figure 3-3 Two capacitors wired in parallel

This experiment proved that a 1000µF electrolytic capacitor retains a charge that is sufficient to keep an LED illuminated for a few seconds—not long enough to use it as a flashlight, but sufficiently long to demonstrate that the "green" flashlight project has potential. The experiment has also shown that if you wire capacitors in parallel, the capacitance increases and the LED remains illuminated longer; wiring two 1000μF capacitors in parallel gave a total capacitance of 2000μF (which is 0.002F). Perhaps, then, wiring capacitors in parallel is the solution for the "green" flashlight. However, assuming that you need a much larger capacitor of, say, 1F to give you the desired results, you would need to wire 1000 capacitors in parallel, which of course would not be very practical for a portable flashlight! So, you need to find a capacitor that has a large enough value to store enough charge to illuminate an LED for a few minutes. Fortunately, since the advent of the "supercapacitor," many options are available.

If you search for supercapacitors, you will find that they come in various shapes and sizes, are very compact, and are rated in farads (not fractions of farads). These capacitors are usually used as memory battery backups (for example, there is probably one in your PC that retains its clock settings while it is switched off). So, now try the 1F capacitor in your circuit. Wire it in place of the 1000μF capacitors, as shown in Figure 3-4, and you are ready to go.

Now connect the battery again for around 10 seconds and then disconnect it. You should find that the LED illuminates for a much longer time period without the battery being connected. In fact, the red LED will start to dim after a few minutes but will remain lit for a much longer period. With that solution at hand, you can start to think about the design of your flashlight.

Project 2
"Green" Pocket LED Flashlight

If you look at the following project specifications, you'll see that you could, in theory, use the breadboard design from the preceding section and build the circuit on stripboard. However, there are some safety features that you also need to incorporate into your circuit, which are outlined after the project specifications. There are also some additional circuit enhancements, described at the end of this chapter.

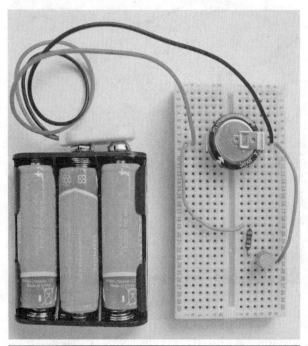

Figure 3-4 1F capacitor breadboard layout. Make sure that the + and − connections of the capacitor are connected to the matching polarities of the battery.

> ### PROJECT SPECIFICATIONS
> - The LED flashlight can be "charged up" by a 9V battery.
> - The LED flashlight is small and portable (pocket size).
> - The LED flashlight uses an internal capacitor to retain its charge.
> - The capacitor can power the bright LED for a few minutes on one charge.

Safety Features

The design of your pocket LED flashlight needs to incorporate the following safety features:

- The maximum voltage for the 1F capacitor used in this circuit is 5.5V, so you need to reduce the 9V supply to below this level.

- You need to protect the circuit against accidentally connecting the 9V supply the wrong way around so that you do not damage the capacitor.

- You need to ensure the capacitor's charge does not discharge if a short-circuit occurs across the battery connections.

How the Circuit Works

Figure 3-5 shows the final circuit diagram for the pocket flashlight, which incorporates all of the safety features outlined in the specification.

The way the circuit works is that when the 9V PP3 battery is applied to the circuit, the current flows through D3, which is a rectifier diode. This component allows current to flow in only one direction, when it is in forward bias. The inclusion of this device in your circuit means that if you accidentally connect the battery the wrong way around, current will not flow through the circuit. This output is then fed into a 5V voltage regulator, REG 1, which converts the voltage into 5V, which is a safe voltage to charge your capacitor, C1. Normally these types of regulators require capacitors to smooth the input and output voltage, but your circuit will work quite well without them, which means that you can also reduce the number of components used in your circuit.

Before the 5V reaches C1, it illuminates the red LED, D1, which is your charge indicator showing that the 9V battery is connected. The 5V then flows through another diode, D4, which allows C1 to be charged and also acts as a buffer to stop C1 from discharging through the circuit when the battery is removed. Finally, SW1 is the on/off switch for the white LED, D2, which provides the light of your flashlight. Pressing SW1 allows the charge from the capacitor to flow through resistor R1 and illuminate the white LED. The value of R1 has been chosen so that less than 1mA flows through D2 (see the notes at the bottom of the parts list).

Parts List

NOTE The Supplier and Part Number column of the following table lists specific parts that I used in this project. Refer to the appendix for additional details about acquiring your parts.

The parts you'll need for the pocket LED flashlight project are listed in the table that follows.

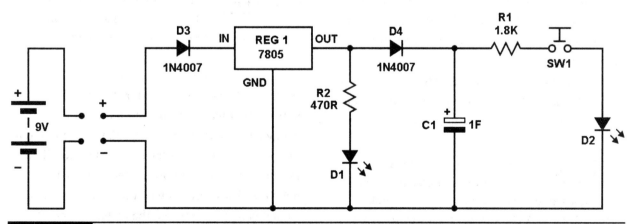

Figure 3-5 "Green" pocket flashlight circuit diagram

PARTS LIST

Code	Quantity	Description	Supplier and Part Number
REG1	1	7805 5V 1A positive voltage regulator	—
D1	1	5mm red LED V_F (typical) = 2.0V, I_F (max) = 30mA	RS Components 228-5972 (pack of 5)
D2	1	5mm high-intensity white LED (6000mcd) V_F (typical) = 3.2V, I_F (max) = 30mA	RS Components 668-6338 (pack of 10)
D3, D4	2	1N4007 1A rectifier diode	—
C1	1	1F 5.5V horizontal electrolytic capacitor	RS Components 339-6843
R1*	1	1.8KΩ 0.5W ±5% tolerance carbon film resistor	—
R2*	1	470Ω 0.5W ±5% tolerance carbon film resistor	—
SW1	1	0.24" × 0.24" (6mm × 6mm) tactile momentary push-to-make switch 0.67" high (17mm), 50mA rated	RS Components 479-1463 (pack of 20)
Hardware	1	Stripboard, 0.1" (2.54mm) hole pitch, 25 holes wide by 9 tracks high (may need to be cut smaller to 23 holes wide; see the text before building the circuit)	—
Hardware	1	PP3 battery clip and lead	RS Components 489-021 (pack of 5)
Hardware	1	9V PP3 battery	—
Hardware	1	Small enclosure (see text)	—
Hardware	—	Double-sided adhesive strip	—

* Note: If you use LEDs that have different V_F and I_F values to those that are in the parts list, then you may need to alter the resistance and wattage values of these LED series resistors. Please refer to Chapter 2, which explains how to do this; also note that R1 has been calculated in this project so that less than 1mA flows through the white LED. This is to ensure that C1's discharge current meets the requirements of the type of capacitor used in the parts list. Because D2 is a high-intensity LED, this is still enough current to provide a decent light output in the dark.

Stripboard Layout

The stripboard layout for this project is shown in Figure 3-6. It is a fairly straightforward design and should not take too long to build. Note that there are four track cuts (represented by white rectangular blocks) that you need to make. Two of the cut tracks will sit beneath the capacitor if you use the same type as the one in the parts list.

NOTE The track cuts that are shown on each of the stripboard layouts in this book are represented by white rectangular blocks. They look like the two track cuts that are shown next to the capacitor in Figure 3-6.

How to Build and Test the Board

NOTE Please refer to Chapter 1 for soldering tips and techniques and for generic stripboard building guidelines.

Now it is time to build the circuit. Begin by soldering the voltage regulator REG1 to the board; notice how I have carefully bent the three component leads so that it lays flat on the board to save space. You should then solder the two diodes D3 and D4, which are positioned near to the regulator. These are shown laid flat on Figure 3-6 so that you can see where the cathode band of the diode should be positioned; however, these two

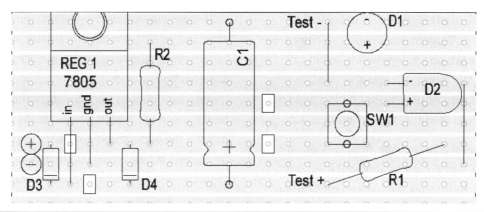

Figure 3-6 "Green" pocket LED flashlight stripboard layout

components should be fitted so that they stand vertically. Once this is done you can complete the build by soldering the two wire links and the other components (except C1 for now) to the board. The last component that you should solder in place is the capacitor, C1. Before you fit this component, you need to perform an electrical test of the circuit to make sure that it works correctly; this will avoid damaging the capacitor. Take a look at the photograph in Figure 3-7. Before fitting the capacitor, connect the 9V PP3 battery to the battery clip, which should illuminate the red charging LED. You then need to use a multimeter to check that the voltage across the two test points is approximately 5V (in reality, the voltage will probably be around 4.5 to 4.7V due to the voltage

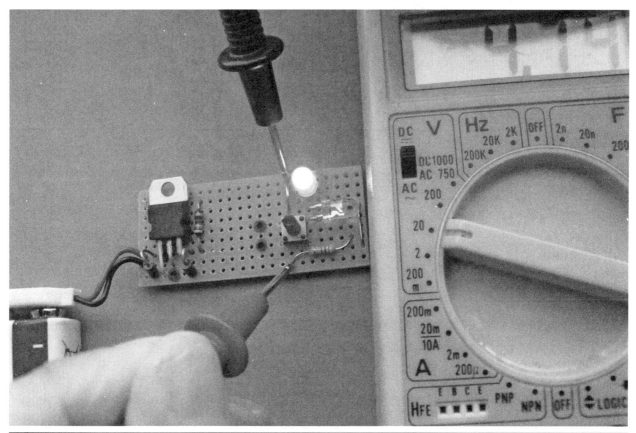

Figure 3-7 Perform the circuit checks before fitting the capacitor.

drop through diode D4). It is also important to check that the polarity of the test point voltages is the same as shown on the stripboard layout in Figure 3-6. If this works as expected, you can now reverse the polarity of the 9V PP3 battery, and the test points should read 0V on your multimeter.

Now connect the battery the right way around again to check that the test points still measure below 5 volts, and if they do you can press SW1 and the white LED should illuminate fairly brightly, as shown in Figure 3-8.

If these checks do not produce the results expected, or if the voltage regulator starts to become hot, then remove the battery immediately and check your circuit, as there is something wrong and you will need to check the board for faults. Once the circuit passes the test, you can remove the battery and solder the capacitor, C1, in place, making sure that its polarity is correct. Figure 3-9 shows the completed stripboard layout of the flashlight.

Now connect the 9V PP3 battery to the circuit for between 30 seconds and a maximum of 3 minutes.

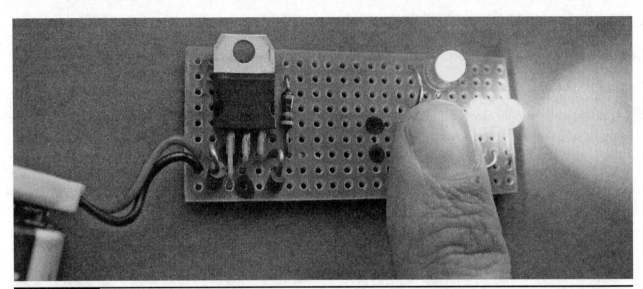

Figure 3-8 Pressing the switch activates the white LED.

Figure 3-9 "Green" pocket flashlight stripboard

The amount of time that the capacitor takes to charge up will depend on how much charge is already stored in C1.

Whenever the 9V battery is connected to the circuit the red LED will be illuminated, showing that the battery is connected and the capacitor will start to charge. After about a minute, remove the battery and press the switch SW1 to see how bright the LED illuminates; if the light output is dim then you will need to charge the capacitor up for a little longer. Once the charging is complete and the battery is removed, the capacitor will retain its charge for a number of weeks. Press SW1, and the white LED will switch on and is bright enough to illuminate a small area in the dark. If you were to keep your finger on the switch, the LED would stay illuminated with a decent light output for around 10 to 15 minutes. With intermittent use of a few seconds at a time, however, the flashlight could remain usable for a few weeks.

To complete your pocket flashlight, you might decide to purchase a small, plastic enclosure in which to mount your stripboard. However, to demonstrate how compact you can make the flashlight, I fitted the stripboard layout into a small, clear-plastic enclosure that originally contained breath freshener mints. I had to create a hole in the side of the box to provide access to the switch, which I did by using the drill bit that I normally use to create breaks in stripboard tracks (that is, I created the hole by hand, not by using an electric drill). The box that I used is ideal for this circuit because it is a clear, flexible enclosure, enabling the white LED to shine through it. I was able to squeeze the box slightly so that I could insert the stripboard and fit the switch through the hole, as shown in Figure 3-10. If you use the same switch that I used, you may also need to pare down the plastic push button of the switch to make it fit in the box. I also had to reduce the width of the stripboard to 23 holes wide before I soldered the components in place, to make sure the stripboard was small enough to fit

Figure 3-10 Squeezing the box to get the electronics inside

inside this box. If you utilize a different enclosure, you may not need to do this.

TIP When building the projects in this book, always compare your completed stripboard with the stripboard layout diagram, the circuit diagram, and the photographs shown in each chapter before you begin testing. This will save you time fault-finding.

I made a small cut in the lid of the box to allow the PP3 battery clip cabling to be fed outside of the box. I then fixed the battery clip to the lid of the box using double-sided tape. The completed project is shown in Figure 3-11.

You have now built a pocket flashlight that you can charge in a matter of seconds and that will provide enough light to illuminate a small area in the dark.

Additional Safety Features

When you are designing electronic circuits, you need to consider the possibility of component failure. The 7805 regulator is a very robust device

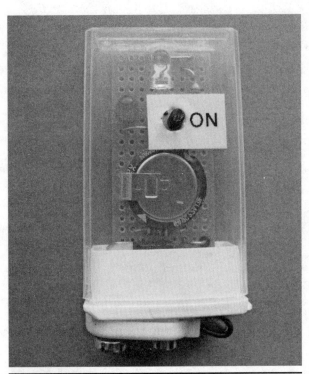

Figure 3-11 The completed flashlight in its unusual enclosure

and should be reliable in this circuit; however, if for some reason (such as mechanical or component failure) the 9V supply bypasses the regulator REG1 to reach the 5.5V-rated capacitor, this could be dangerous. If you find that when you charge up the flashlight, that the capacitor becomes hot, or if it starts to leak or smell, you must remove the battery immediately and find out where the problem lies.

You may therefore want to consider experimenting and designing some additional safety features into the prototype, such as a secondary back-up voltage regulator or a circuit to cut out the 9V supply in case of a problem. Or alternatively, you may even decide to modify the circuit so that you only charge up the flashlight with a 4.5V supply (3 × AA batteries) instead.

In version 2 of my prototype, I decided to add an additional warning LED indicator that illuminates only when the voltage heading to the capacitor exceeds 5.5 volts. If this LED ever illuminates when the flashlight is being charged up, it would prompt you to disconnect the battery from the circuit immediately. Take a look at Figure 3-12, which shows how I modified the circuit to include a 5.6V Zener diode (1.3-watt rated), an additional red LED, and a 1KΩ resistor (1-watt rated) to create this feature.

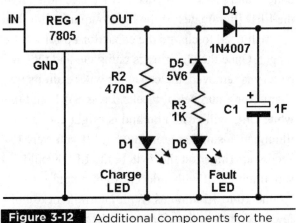

Figure 3-12 Additional components for the warning LED, which are fitted after the regulator

TIP Try to charge the flashlight when the white LED becomes dim and not when it no longer illuminates. Charging the flashlight before the capacitor becomes fully discharged will reduce the amount of current that needs to be drawn from the PP3 battery to charge up the flashlight, thus increasing the life of the battery.

Possible Circuit Modifications

You might want to experiment to see if you can alter the circuit slightly. For example, could you modify the circuit so that the red charging LED switches off when the capacitor is fully charged? You might also want to make the circuit a true "green" eco-friendly flashlight by using an alternative renewable supply rather than the 9V PP3 battery. One idea is to see if it is possible to modify the circuit so that you can use a solar panel to charge the capacitor.

CHAPTER 4

Building a Clock Generator: Basic Single-LED Flasher

IF YOU COMPLETED THE FIRST TWO PROJECTS IN this book, you now have mastered the art of illuminating an LED and are ready to move on to making an LED flash on and off. To create a flashing effect, you need to create a clock generator circuit. There are numerous circuit building blocks that you could use to do this, and you will discover some of them as you work through this book, starting in this chapter. *Clock generators* are circuits that produce a regular and never-ending train of digital pulses that alternates between a low state and a high state. If you were to connect an oscilloscope to the output of a clock circuit, you would see a waveform similar to that shown in Figure 4-1.

We are now going to discover a popular clock generator integrated circuit (IC) which has been available to the electronics hobbyist for many years, this being the 555 timer.

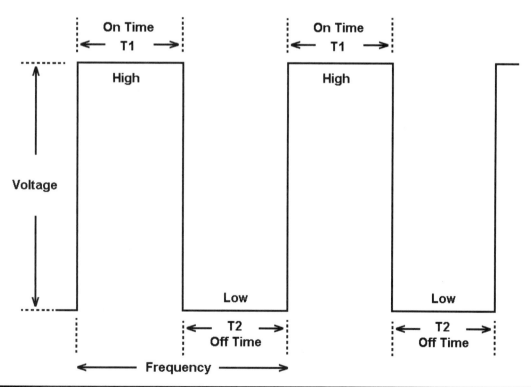

Figure 4-1 Digital clock pulse diagram

41

The 555 Timer

Clock circuits are also sometimes known as astable circuits, oscillators, or multivibrators. One of the easiest methods of producing a clock generator is to use the popular 555 timer chip, which is a cost-effective 8-pin device that requires only two resistors and two capacitors to function. The 555 timer has two modes of operation, called monostable and astable modes. The monostable mode allows you to create a one-shot time delay (a single on/off pulse); you will not be using this mode in any of the projects in this book. Instead, you will be using the astable mode, which is the operational state required to flash an LED on and off. The circuit diagram for the 555 timer configured in astable mode is shown in Figure 4-2.

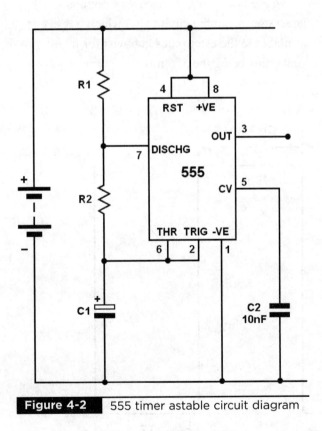

Figure 4-2 555 timer astable circuit diagram

Pin 3 of the 555 timer is the output pin, and this is where the clock signals are generated. Later in the chapter, you will see how you can use this output to flash an LED on and off. The output timing sequence of the 555 timer is configured by altering the values of R1, R2, and C1.

555 Timer Variants

There are a few versions of the 555 timer available, including a low-power version (sometimes known as the 7555 timer), and their specifications do vary slightly, including their supply voltage and output current capabilities. Also, some 555 timers require a capacitor fitted between pin 5 and the negative supply rail, while some do not need it. There is also a dual version of the 555 timer, which is available in a 14-pin package and contains two timers in a single device; this is called the 556 timer. Keep this version in mind for projects that require more than one clock generator, as it will save you from having to use two separate ICs in your circuit designs.

555 Astable Timing Formulas

Although I would prefer to avoid the use of heavy mathematics in this book, it is important for you to understand several useful formulas that you can use to calculate the clock timings of the astable output of the 555 timer. These formulas are outlined next.

NOTE In addition to the timing calculation examples in the following section, Table 4-1, later in the chapter, provides example component values for R1 and R2 and C1 that you can use to create useful output timings to pulse an LED on and off.

- *Charge time* (T1) of the timer (high output) is calculated by using this formula:

  $T1 = 0.693 \times (R1 + R2) \times C1$

- *Discharge time* (T2) of the timer (low output) is calculated by using this formula:

  $T2 = 0.693 \times R2 \times C1$

- *Total time period* (T), which is the total of the on cycle and off cycle, is therefore:

  T = T1 + T2

  or

  T = 0.693 × (R1 + (2 × R2)) × C1

- *Frequency of oscillation* (F) equals the number of times T1 + T2 occurs every second and is expressed in hertz (Hz):

  F = 1 / T

  or

  F = 1.44 / ((R1 + (2 × R2)) × C1)

- *Mark (on time) to space ratio (off time)* is calculated as follows:

  (R1+R2) / R2

NOTE For example, a mark-to-space ratio of 3 means that the output is on for three times as long as it is off. A mark-to-space ratio of 1 means that the on and off times are the same. To create a roughly equal mark-to-space ratio, the value of R2 needs to be a lot higher than the value of R1.

Example 555 Astable Timing Calculations

To give you concrete examples of how the formulas provided in the preceding section work, this section inserts the following three example resistor and capacitor values into the formulas and shows you the results (note that R2 is a lot higher than R1). Bear in mind that the tolerance of these components will vary, so in reality these timings may not be exact.

R1 = 12,000Ω

R2 = 180,000Ω

C1 = 4.7µF

We need to make sure that the resistor values used in the formulas are in ohms (Ω) and that the capacitor value is converted to farads (F); this is why a value of 0.0000047F is used in the formula for the 4.7µF capacitor.

- *Charge time* (T1) of the timer (high output):

  T1 = 0.693 × (12000 + 180000) × 0.0000047 = 0.63 sec

- *Discharge time* (T2) of the timer (low output):

  T2 = 0.693 × 180000 × 0.0000047 = 0.59 sec

- *Total time period* (T; on and off cycle):

  T = 0.63 sec + 0.59 sec = 1.22 sec

  or

  T = 0.693 × (12000 + (2 × 180000)) × 0.0000047 = 1.21 sec

- *Frequency of oscillation* (F):

  F = 1 / 1.21 = 0.82 Hz

  or

  F = 1.44 / ((12000 + (2 × 180000)) × 0.0000047) = 0.82 Hz

- *Mark-to-space ratio*:

  (12000 + 180000) / 180000 = 1.07

So in this example, the on and off times are almost the same, which is confirmed by the mark-to-space ratio being almost exactly 1. This is achieved by the value of R2 being a lot higher than R1.

The recommended range of component values to use with the 555 timer, for accuracy, is between 1KΩ and 1MΩ for R1 and R2, and for flashing LED circuits you should not need to use a capacitor value of more than 100µF for C1.

Source Current and Sink Current

If you chose to build the LED flashlight in Chapter 2 (project 1), you learned that one of the key considerations required when building LED circuits is to understand the total drive current (or *source current*) available in the circuit to illuminate an LED. When using the 555 timer, this drive current is determined by the total output current limitations of pin 3. The datasheet for a 555 timer expresses the output current available from output pin 3 in milliamps (mA). Depending on the type of 555 timer that you use, the output current is normally 100mA or 200mA, which means that this chip can drive and illuminate quite a few LEDs, assuming that a single LED might draw around 15mA to 20mA.

The term *sink current* is also important to understand, so here is a more detailed explanation of these two terms in the context of the following project:

- *Source current* When the positive (high) clock cycle is used to drive an LED. In this situation, pin 3 would be connected to the anode side of an LED, and the cathode side would be connected to the negative rail of the circuit. You are basically using pin 3 as a positive rail to drive the LED.

- *Sink current* When the negative (low) clock cycle is used to drive an LED. In this situation, pin 3 would be connected to the cathode side of an LED, and the anode side would be connected to the positive rail of the circuit. You are basically using pin 3 as a negative rail to drive the LED. You should note that not all ICs are suitable for sinking current in this way.

Example source and sink current layout configurations for the 555 timer are shown in Figure 4-3.

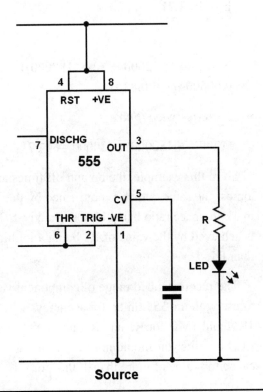
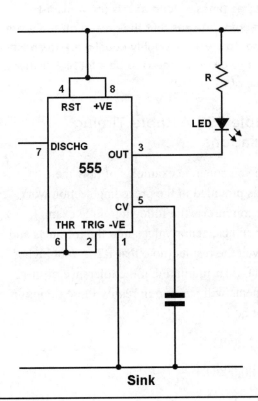

Figure 4-3 Example source and sink current configurations

Project 3
Basic Single-LED Flasher

In this project, you'll build an experimental single-LED flasher circuit on a piece of stripboard using the 555 timer.

> **PROJECT SPECIFICATIONS**
>
> - The experimental 555 timer board shows you how to flash an LED on and off.
> - The LED flashing rate can be altered by changing the resistor and capacitor values.
> - The timing components can be easily changed without having to desolder them from the board.

How the Circuit Works

The diagram in Figure 4-4 shows the circuit layout required to build a simple flashing LED circuit. Note that it is similar to the circuit layout shown in Figure 4-2 earlier in the chapter.

As you can see, you are going to use the source current of pin 3 to flash the LED. Because the 555 timer has a good source current capability, there is no requirement to boost the current output; pin 3 can easily drive the LED directly. The 555 timer that I used for my circuit has a wide supply voltage range of between 4.5V and 16V, and in this circuit I used a 6V supply provided by four AAA batteries. As this stripboard is intended for experimentation, the values of R1, R2, and C1 are not shown in the circuit diagram, and you will not actually solder R1, R2, and C1 onto the stripboard in this project. Instead, you'll use SIL turned pin sockets, which allow you to experiment with the values of these three components without having to solder them on and then desolder them from the stripboard. The SIL sockets are normally supplied in strips of 20 sockets, but you can cut these to size with your wire cutters, making sure that you cut through only the soft plastic shell of the SIL socket.

A variable resistor (VR1) is included to allow you to further adjust the speed of the clock pulses. This is a 1MΩ variable resistor, which allows you to add another 0Ω to 1MΩ to the value of R1. Don't forget that the accuracy of the timer will suffer when the value of R1 or R2 is above 1MΩ.

Parts List

> **NOTE** The Supplier and Part Number column of the following table lists specific parts that I used in this project. Refer to the appendix for additional details about acquiring your parts.

The parts you'll need for the basic single-LED flasher project are listed next.

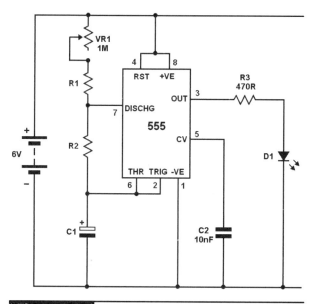

Figure 4-4 LED flasher circuit diagram

PARTS LIST

Code	Quantity	Description	Supplier and Part Number
IC1	1	555 timer chip	RS Components 534-3469 (or similar)
D1	1	5mm red LED V_F (typical) = 2.0V, I_F (max) = 30mA	RS Components 228-5972 (pack of 5)
C2	1	10nF ceramic disk capacitor (min. 10V rated)	—
R3*	1	470Ω 0.5W ±5% carbon film resistor	—
VR1	1	1MΩ miniature enclosed horizontal preset potentiometer (min. 0.15W rated)	—
Hardware	1	Stripboard, 0.1" (2.54mm) hole pitch, 25 holes wide by 9 tracks high	—
Hardware	1	20-way turned pin SIL socket strip (cut to size)	RS Components 267-7400 (pack of 5)
Hardware	1	AAA battery holder (four AAA batteries)	RS Components 512-3568
Hardware	4	AAA battery (1.5V)	—
Hardware	1	8-pin DIL socket	—
R1, R2	—	Various resistor values (0.5W carbon film)	—
C1	—	Various capacitor values (min. 10V rated)	—

*Note: If you use an LED that has different V_F and I_F values to the one specified in the parts list, then you may need to alter the resistance and wattage values of this LED series resistor. Please refer to Chapter 2, which explains how to do this.

Stripboard Layout

The circuit is built on a small piece of stripboard, the layout of which is shown in Figure 4-5. Note that you need to cut five tracks, one near to VR1 and the other four under the 555 timer DIL socket. Also note that two of the wire links almost wrap around the 555 timer; one wire link connects pins 4 to 8 together and the other links pins 2 to 6 together. The SIL sockets are shown as large, square blocks on the diagram.

How to Build and Test the Board

NOTE Please refer to Chapter 1 for soldering tips and techniques and for generic stripboard building guidelines.

Build the stripboard layout based on Figure 4-5 (after cutting the five tracks), and solder the wire links, capacitor (C2), and sockets in place. You should end up with a stripboard layout that looks similar to Figure 4-6.

Chapter 4 ■ Building a Clock Generator: Basic Single-LED Flasher   47

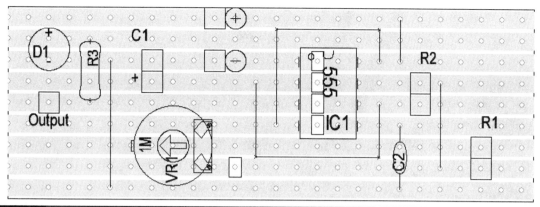

Figure 4-5   LED flasher stripboard layout

Figure 4-6   The completed stripboard layout

After you complete the stripboard, make sure that the component layout matches Figures 4-5 and 4-6 and that there are no solder bridges on the trackside of the board. Then, fit the resistor and capacitor components into the SIL sockets, as shown in Figure 4-7, using the same component values derived from the calculations in the "Example 555 Astable Timing Calculations" section earlier in the chapter.

R1 = 12,000Ω

R2 = 180,000Ω

C1 = 4.7µF

Make sure both that the capacitor is inserted the correct way around, as it is an electrolytic type, and that the variable resistor is wound fully clockwise (this means that the variable resistor has zero resistance). Also make sure that none of the component leads are touching each other or any other part of the circuit. Now apply the battery to the circuit, and the LED should flash in a steady manner. If you turn the variable resistor slowly counterclockwise, the speed of the flash should

Figure 4-7 Stripboard with the components inserted into the SIL sockets and the battery leads connected

become slower because you are increasing the resistance of R1.

By experimenting with the values of the three components, you can see the various results that you can achieve. Don't forget to remove the power to the board before you change the values of the components. This experimental board allows you to study the effects of altering the component values on the flashing speed of the LED. Keep hold of the board because you will be using it again in Chapter 17 to test the digital oscilloscope screen.

Table 4-1 shows some example resistor and capacitor values for you to try out on the LED flasher board (the effects are based on VR1 being wound fully clockwise).

TABLE 4-1	Useful Astable Timings for the Flasher Board							
R1	R2	C1	On (sec)	Off (sec)	Freq (Hz)	Total Time (sec)	Mark/ Space Ratio	Effect
12KΩ	180KΩ	4.7μF	0.63	0.59	0.83	1.21	1.07	Steady flash
12KΩ	180KΩ	1μF	0.13	0.12	3.88	0.26	1.07	Faster flash
220KΩ	220KΩ	10μF	3.05	1.52	0.22	4.57	2.00	Slow flash
47KΩ	470KΩ	10μF	3.58	3.26	0.15	6.84	1.10	Slower flash
10KΩ	100KΩ	1μF	0.08	0.07	6.87	0.15	1.10	Fast flash
39KΩ	39KΩ	1μF	0.05	0.03	12.33	0.08	2.00	Fast strobe
82KΩ	82KΩ	1μF	0.11	0.06	5.87	0.17	2.00	Slower strobe

Experiments to Try on Your Own

Experiment with different component values to see what results you get and try to understand what occurs to the flashing timings by increasing or decreasing their values. Also try out nonelectrolytic capacitors to see the effect on the LED. Remember that if the LED does not flash but remains illuminated, it might actually be flashing too quickly to be seen by the naked eye. If you have an oscilloscope, try connecting the positive lead of the scope to output pin 3 of the IC and connecting the ground lead to the negative pin 1; you should then see a square wave output similar to the one shown earlier in Figure 4-1, enabling you to measure its frequency.

CHAPTER 5

Bringing a 555 Timer to Life: LED Bike Flasher

CHAPTER 4 DESCRIBED HOW TO CREATE AN experimental clock circuit that is capable of flashing a single LED on and off using a 555 timer. If you completed that project (project 3) and experimented with various resistor and capacitor values, you understand how you can modify the flashing sequence by altering these three component values. In the project in this chapter, you are going to use the astable circuitry from project 3 to create a flashing LED circuit that you can mount on your bicycle for use as a visual warning device or as illumination in the dark. An example of the completed project is shown in Figure 5-1.

CAUTION If you decide to install the flasher from this project on your bicycle, please observe your local legal requirements relating to bike illumination and ensure that flashing LEDs are acceptable for use on public streets and roads. Also note that this project is meant as supplementary illumination for your bicycle and should be used in conjunction with your normal front and rear bike lights.

The 555 timer used in project 3 has source and sink current capabilities of around 200mA, which is plenty to drive a single LED and more besides. Figure 4-3 in the previous chapter shows how you can configure the 555 timer to either source current or sink current to an LED by connecting the LED to either the positive or negative rail. These configurations will be used next as the basis for project 4.

Project 4
LED Bike Flasher

There are two parts to this project, "Rear Red Flasher" and "Front LED Flasher," and it is up to you whether you want to build the core design only (rear flasher) or build the complete project (both rear and front flashers).

Figure 5-1 LED bike flasher

PROJECT SPECIFICATIONS

- The flashing LED circuit is capable of driving multiple LEDs.
- The flashing sequence is eye catching and helps to make you and your bike more visible in the dark.
- The core design for the bike flasher has four red LEDs to provide illumination at the rear of your bike.
- The second, optional part of the project shows you how to connect an additional two white LEDs to the circuit, providing illumination at the front of your bike.
- The circuit is compact enough to enable it to be housed in a small enclosure that can be mounted underneath the seat of a bicycle.
- The supply voltage is 9 volts.
- The current consumption of the flasher is fairly low, which enables many hours of use from a single 9V battery.

How the Rear LED Flasher Circuit Works

Figure 5-2 shows the circuit diagram for the project. If you completed project 3, the circuit should look familiar because it uses the 555 timer in astable mode, with the values of R1, R2, and C1 set to provide a fast flash sequence that pulses at around five times per second. Using the timing formulas from Chapter 4, the timing sequence of this circuit is shown below.

$R1 = 10K\Omega$

$R2 = 150K\Omega$

$C1 = 1\mu F$

- *Charge time* (T1) of the timer (high output):

  $T1 = 0.693 \times (10000 + 150000) \times 0.000001$
  $= 0.11$ sec

- *Discharge time* (T2) of the timer (low output):

  $T2 = 0.693 \times 150000 \times 0.000001 = 0.10$ sec

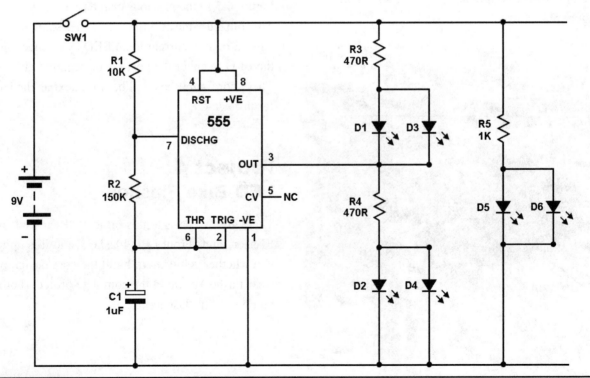

Figure 5-2 Bike flasher circuit diagram

- *Total time period* (T; on and off cycle):

  T = 0.11 sec + 0.10 sec = 0.21 sec

- *Frequency of oscillation* (F):

  F = 1 / 0.21 = 4.76 Hz

- *Mark-to-space ratio:*

  (10000 + 150000) / 150000 = 1.07

In this project, you will use a 9V battery to power the circuit. Although the amp-hour (Ah) capacity of a 9V PP3 battery is typically lower than AAA or AA cells, the PP3 battery is quite compact, which is ideal for this project, and if you make the power consumption of the circuit low enough, you can still get many hours of use from the PP3 cell.

The circuit operates in a similar manner to the circuit in the single-LED flasher project from Chapter 4, but, as you can see in Figure 5-1, there are now four red LEDs, which are wired up in pairs (D1 to D4). Two of the LEDs are wired so that the 555 timer output pin 3 sources current (D2 and D4) and the other two LEDs sink current through the timer (D1 and D3). When the main switch (SW1) is switched on, the timer (IC1) is powered by the 9V battery and begins to operate in astable mode. When its output pin is high, it illuminates only two of the LEDs (D2 and D4), and when the output switches to a low output, these LEDs turn off and the other two LEDs (D1 and D3) illuminate. This means that the LEDs provide an alternating flashing sequence. Because the speed of the astable is fairly fast, the alternating flashing effect is very eye catching. (Police and emergency service vehicles use this type of flashing effect for the same reason.)

While LEDs D1 to D4 form the core circuit for the rear flashing LEDs for the bike, you also have the option to build a front illuminator that comprises two white LEDs (D5 and D6). These two LEDs do not flash but provide a fairly good light output when the circuit is switched on. The value of the resistor that feeds these two LEDs is set at 1KΩ, which means that the combined current draw for both white LEDs is reduced to around 12mA. However, because these are the same type of high-intensity LEDs that were used previously in the flashlight projects, they still provide a decent enough light output even when the current flowing through them is this low.

NOTE Each pair of LEDs in the circuit is driven by a single series resistor, which is not always recommended but worked perfectly fine in my prototype. Depending on the properties of the LEDs that you use, this configuration may not provide an even illumination for each pair of LEDs. If this is the case, then you may need to build a slightly modified circuit, as discussed at the end of this chapter.

Parts List

NOTE The Supplier and Part Number column of the following table lists specific parts that I used in this project. Refer to the appendix for additional details about acquiring your parts.

The parts you'll need for the LED bike flasher project (optional parts for the front LED flasher are indicated in parentheses) are listed next.

PARTS LIST

Code	Quantity	Description	Supplier and Part Number
IC1	1	555 timer chip	RS Components 534-3469 (or similar)
D1–D4	4	5mm red LED V_F (typical) = 2.0V, I_F (max) = 30mA	RS Components 228-5972 (pack of 5)
R1	1	10KΩ 0.5W ±5% tolerance carbon film resistor	—
R2	1	150KΩ 0.5W ±5% tolerance carbon film resistor	—
C1	1	1μF 16V electrolytic capacitor	—
R3, R4*	2	470Ω 1W ±5% tolerance carbon film resistor (see text)	—
(R5)*	1	1KΩ 1W ±5% tolerance carbon film resistor (see text)	—
(D5, D6)	2	5mm high-intensity white LED V_F (typical) = 3.2V, I_F (max) = 30mA	RS Components 668-6338 (pack of 10)
SW1	1	Toggle switch single pole single throw (SPST), 2A rated	RS Components 710-9671
Hardware	1	Stripboard, 0.1" (2.54mm) hole pitch, 25 holes wide by 9 tracks high	—
Hardware	1	PP3 battery clip and lead	RS Components 489-021 (pack of 5)
Hardware	1	9V PP3 battery	—
Hardware	1	8-pin DIL socket	—
Hardware	2	LED clips (to suit 5mm LEDs for the front LED flasher only; see text)	—
Hardware	—	Enclosures (see text)	—
Hardware	—	Cable, cable tie bases, and cable ties (for front LED flasher only; see text)	—

*Note: If you use LEDs that have different V_F and I_F values to those specified in the parts list, then you may need to alter the resistance and wattage values of these LED series resistors. Please refer to Chapter 2, which explains how to do this.

Stripboard Layout

The stripboard layout for the LED bike flasher is shown in Figure 5-3. Some 555 timers require a capacitor between pin 5 and the negative battery connection, but I did not include one in my build and the circuit seemed to work fine. If you decide to include a capacitor at that location, you will need to add it to the parts list and solder it onto the stripboard layout accordingly. You need to cut four tracks on this layout, and these sit beneath the DIL socket that the timer chip plugs into.

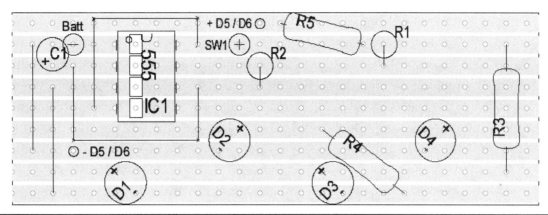

Figure 5-3 Stripboard layout for the rear LED flasher

How to Build the Board

NOTE Please refer to Chapter 1 for soldering tips and techniques and for generic stripboard building guidelines.

Build the stripboard layout based on Figure 5-3. You should end up with a stripboard that looks like the one shown in Figure 5-4. The four LEDs are connected in alternate pairs, which means that D1 and D3 illuminate together and D2 and D4 illuminate together, as shown in Figure 5-5, which helps to provide the eye-catching flashing sequence. If you decide that you want to build the front illuminator as well, then you will need to include resistor R5 on the stripboard.

I soldered the four LEDs in such a way that they lay down flat on the stripboard. I did this so that

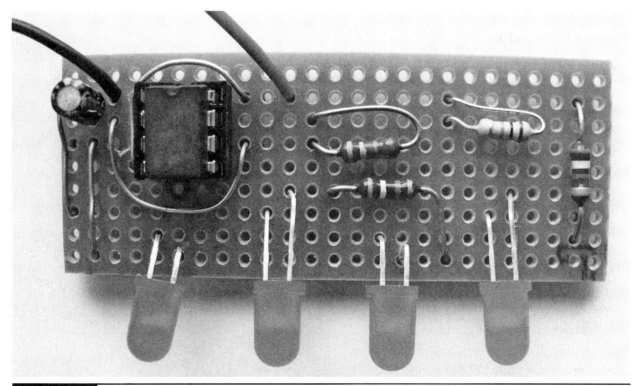

Figure 5-4 The rear flasher stripboard

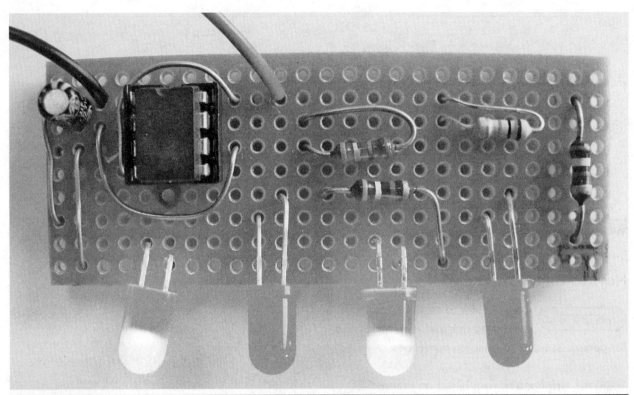

Figure 5-5 Completed stripboard with the LEDs flashing

the board would then fit neatly into the enclosure that I used for the project (more on this shortly).

Front LED Flasher

If you decide to build the complete project with the two additional LEDs for the front of your bike, you can connect the white LEDs to the stripboard circuit by connecting a suitable length of two-core cable to the points outlined on the stripboard layout in Figure 5-3. You also need to solder resistor R5 onto the stripboard. Whenever I build electronic projects, I am always trying to use components or enclosures from my hoard of old salvaged parts (I like to recycle, plus it saves me some money along the way). I therefore decided to take an unconventional approach and save some money on an enclosure by using an old dental floss container, which I drilled three holes into and painted black. It is very compact and is ideal for mounting the two white LEDs. You may decide to purchase a standard small, off-the-shelf enclosure for your project. Figure 5-6 shows the unusual enclosure I used with the two white LEDs mounted

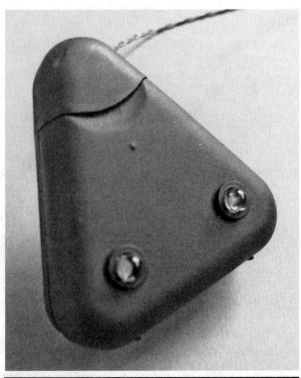

Figure 5-6 Front LED enclosure

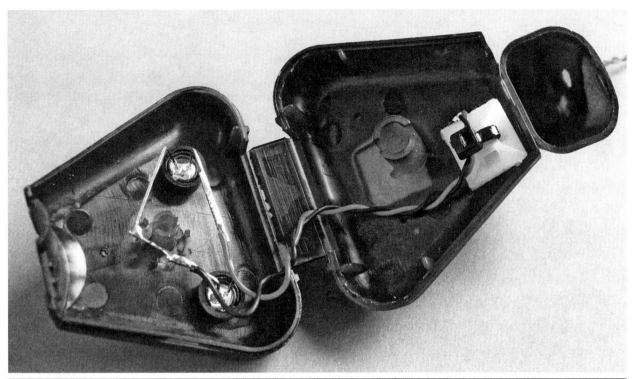

Figure 5-7 Solder the two LEDs together and connect them to the cable.

on the front, which are fitted into LED clips. Figure 5-7 shows the inside of the box and the two white LEDs soldered together (anode to anode and cathode to cathode). These are then soldered to a length of two-core cable, which is then fed out of the box and secured in place with a cable tie pad and a cable tie.

Enclosure for the LED Bike Flasher

For the main control unit, I used a salvaged enclosure that used to house the electronics and infrared LEDs for a wireless headphone transmitter. It broke down a while ago, and I was glad that I kept the enclosure after I dismantled it because the enclosure is very compact and the stripboard and red LED display configuration fit into it nicely. You may decide to utilize your own salvaged enclosure or purchase a suitable enclosure from one of the component suppliers outlined in the appendix.

Figure 5-8 shows the complete project with the main stripboard mounted inside the enclosure and the toggle power switch mounted on the back of the box. The positive lead of the PP3 battery clip is soldered directly to the switch, and the other side of the switch is then wired to the stripboard. The negative battery clip lead is soldered directly to the stripboard. The 9V PP3 battery fit snugly in the space at the top of my enclosure.

Finally, I attached two cable tie pads securely onto the other half of the enclosure using small screws. I used these cable tie pads as fixing points to mount the complete enclosure underneath the seat of the bike, as shown in Figure 5-9. My bike seat has two lugs underneath that are included for attaching accessories, and these were ideal for securing the enclosure to the seat, which I did by using two cable ties.

58 Brilliant LED Projects

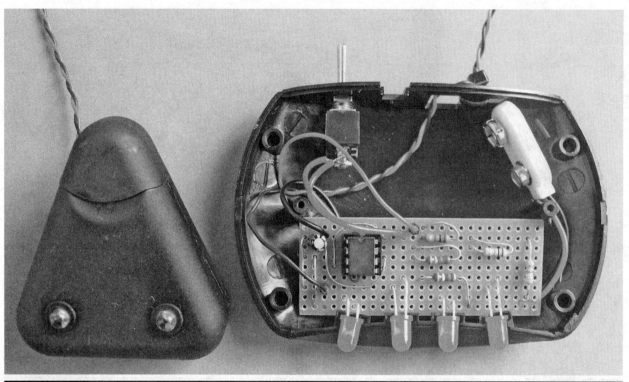

Figure 5-8 The completed LED bike flasher project

Figure 5-9 The rear LED flasher mounted underneath the bike seat

Figure 5-10 The front LED illuminator mounted on the bike

Figures 5-10 and 5-11 show how I mounted the two LED enclosures onto my bicycle. The interconnecting cable runs from the back of the bike from the stripboard module to the white LED module at the front of the bike. I made sure that the interconnecting cable was fixed securely to the frame of the bike using cable ties to ensure that the cable is not loose and does not get tangled or trapped in the wheels or chain. Both enclosures were secured to the bike using cable tie bases and cable ties, making sure that the cable tie bases were securely fixed to the enclosures.

Figure 5-11 The rear LED flasher illuminated

One thing to note when you mount the front and rear enclosures on your bike is that they are exposed to the elements, so you need to ensure that they are suitably protected from the rain and puddles. The rear enclosure is in a fairly good position because it is partially protected by the seat, but it will still get splashed from time to time. I suggest using some silicone sealant around any gaps in the enclosures to help protect against water ingress.

Now all you have to do is flick the switch on, which is mounted on the rear enclosure beneath your seat, and both the front and rear LEDs activate. The total current consumption of the unit with all six LEDs illuminated is only around 25–30mA, which means that the battery will work continuously for quite a few hours before it needs to be replaced.

Experimenting to Reduce the Current Consumption

When I built this project, I also experimented with the circuit by trying out a 7555 timer chip (a low-power version of the 555 timer) to see what would happen. I wanted to see if this would reduce the overall power consumption of the unit. Normally, a standard 555 timer might draw between 5–10mA from the battery without any LEDs connected; this might not sound like a lot, but anything we can do to reduce this will make a big difference in how long the battery will last. The low-power 7555 version draws only a few hundred microamps (μA), which is a negligible current draw and means that the bulk of the circuit current consumption will be taken up by the LEDs.

While experimenting, I realized that the output current capability of the 7555 timer is 100mA (as opposed to 200mA for the 555 version), and I was probably also pushing the boundaries of its current sink capabilities. However, the 7555 timer chip did seem to work okay when I experimented with it. When I used the 7555 timer, the total current consumption of the unit was reduced to around 20mA, but the alternating switching of the red LEDs seemed to look visibly "cleaner" when I used the 555 timer rather than the 7555 version. Although I've tested the circuit for a number of hours with the 7555 timer, I am not sure what the long-term effects of using this version would be in this circuit.

Another experiment that you might want to try out is to see if it is possible to power the LED Bike Flasher using peddle power. One way to do this would be to use a dynamo which generates electricity when you peddle the bike; you could probably buy one of these from a bike shop or you could make one using a low voltage DC motor. If you decide to try this out, your circuit design needs to incorporate some method of voltage rectification, smoothing, and regulation to ensure that the circuit receives a clean voltage supply (and at the correct voltage for the circuit). You will also need to remember that the supply voltage will stop when your bike stops, which will mean that the LEDs will go off when you come to a stop. You will therefore need to think about incorporating a battery back-up circuit which will switch on and power the LED Bike Flasher when you stop at traffic lights or road junctions.

Alternative Circuit

If you find that powering each pair of LEDs through a single series resistor doesn't work very well in your project, you may need to create a slightly modified circuit diagram and stripboard layout. You would need to incorporate three additional resistors (two more 470Ω resistors, one each for D3 and D4, and one more 1KΩ resistor for D6) and three additional track cuts to allow each LED to be fed by its own series resistor. If you build this alternative circuit, you will need to make sure that the two anodes (+) of the front white LEDs (D5 and D6) are not soldered together as described earlier in the chapter. Also, you probably need a three-core cable instead of a two-core cable to link the control unit to the front LEDs.

CHAPTER 6

Exploring Multicolor LEDs: Color-Changing Light Box

THE PROJECTS IN THE PRECEDING CHAPTERS INVOLVED only single-color LEDs, either red or white. This chapter shows you how to add variety to the colors you use in your projects. There are three primary types of LED available that contain more than one color in a single device. These multicolor LEDs are quite flexible and will allow you to create some interesting effects in your projects. This chapter first explores each of these types of LED and then shows you how to build a color-changing light box similar to the one shown in Figure 6-1. This can be used, for example, as mood lighting or to provide a mini-disco lighting effect.

Figure 6-1   Color-changing light box

Multicolor LEDs

Three main types of multicolor LEDs are available, as described next. These types of LED contain two or three different color elements in the same package, allowing you to create some interesting colorful effects in your projects.

Bicolor LEDs

This type of LED has two legs, like its single-color counterpart, but unlike a monochrome LED, it will illuminate when it is reverse biased. You still require a series resistor to protect the LED as normal, but the difference with this device is that if you connect the LED across the battery in one direction, the color of illumination will be red, and if you connect the LED in the opposite direction, then the color output changes to green. The nice thing about this device is that you have only two leads to configure to create two different colors. The downside is that you can display only one color at a time, although you can alternate the supply voltage to the LED at a fast enough speed that you can fool the eye into seeing a yellow or amber color.

A typical bicolor LED is shown in Figure 6-2. If the shorter lead, which is positioned next to the flat side of the package, is connected to the positive (+) side of the battery, and the longer lead is connected to the negative (−) side of the battery,

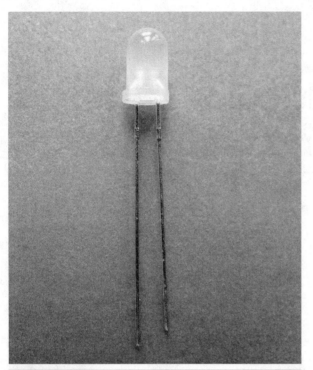

Figure 6-2 Bicolor LED

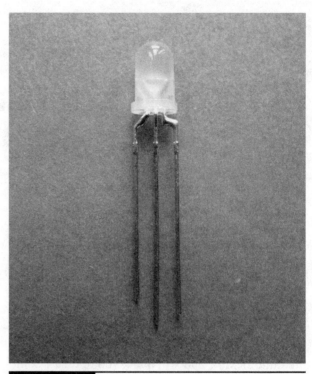

Figure 6-3 Tricolor LED

the LED will glow green. If the battery leads are reversed, the LED will glow red instead. Do not forget that you still need to use a series resistor with these LEDs.

Tricolor LEDs

This type of LED has three leads, normally with two anode (+) connections and a single common cathode (−) connection. This LED also contains two color elements, red and green. The difference between this type of LED and the bicolor LED is that three colors can be generated without having to alternate the supply voltage. If power is applied to a single anode lead, the output color will be either red or green depending on which anode is activated. If both anodes are activated at the same time, then both the red and green parts of the LED illuminate, which fools the eye into seeing a yellow or amber color.

This device is great to use in digital circuits because you can control the three output colors very easily. A tricolor LED is shown in Figure 6-3. The middle connection for the device is a common cathode (−) connection for both colors, the lead next to the flat side of the package is the red anode (+) connection, and the other lead, on the left side, is the green anode (+) connection.

Red, Green, and Blue (RGB) LEDs

The third main type of LED is the RGB LED, which contains three different color emitters in a single package, these being red, green, and blue. The RGB LED works in a similar manner to the tricolor LED, enabling you to create multiple colors depending on the combination of anode connections that is activated. Because there are three colors in the LED, there are actually seven different ways that you can activate these leads, and therefore there are seven different color combinations that can be generated by the single LED, depending on the way it is wired up. In fact, this type of LED is much more versatile and can create a palette of colors that is much more diverse than seven colors, depending on how bright you illuminate each of the color emitters in the LED.

The RGB LED uses the same principle that is used in a color TV screen, where each screen pixel contains three color filters (RGB) and each of the colors is adjusted in intensity to provide the required color output. For example, if red and blue are switched on at the same time, we see this combination as a purple color. A photograph of an RGB LED is shown in Figure 6-4. There are various pin configurations for these types of LEDs, so you should always refer to the LED datasheet.

Bi-Color LED

Tri-Color LED

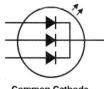
Common Cathode RGB LED

Figure 6-5 Multicolor LED schematic symbols

Project 5
Color-Changing Light Box

The purpose of this project is to show you how to use an RBG LED to create various colors. It is a nice novelty project that you can use as the basis for other mood-lighting or disco-effect projects.

PROJECT SPECIFICATIONS
■ The light box is a compact unit that uses a single RGB LED.
■ The light box slowly cycles through seven different color combinations.
■ The speed at which the colors change can be varied.

As previously mentioned, changing the input configurations to an RGB LED allows you to create various color combinations, and this forms the basis of this project. You therefore need to build some sort of clock circuit that has three outputs that can be used to activate the three individual colors of the LED in various combinations. Using three individual 555 timers to create a visual effect is one option, but I decided to use a different circuit concept instead.

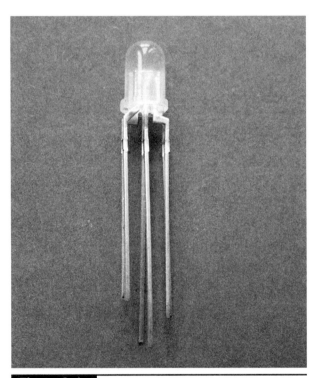

Figure 6-4 RGB LED

Symbols for Multicolor LEDs

Typical schematic symbols used for the three different types of multicolor LED are shown in Figure 6-5.

How the Circuit Works

The circuit diagram for the color-changing light box is shown in Figure 6-6.

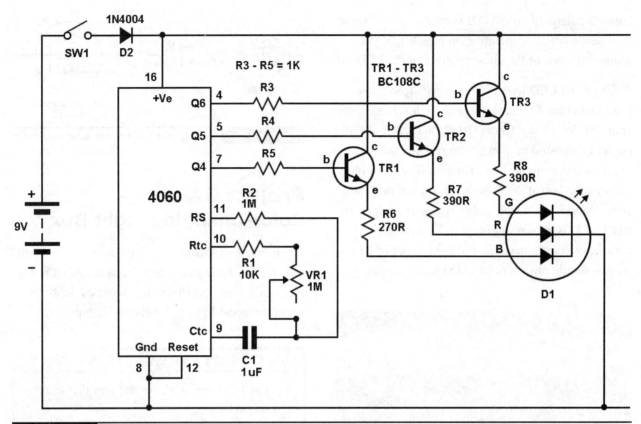

Figure 6-6 Circuit diagram for the color-changing light box

As the circuit diagram shows, I decided to use a 4060 chip at the heart of the circuit. The 4060 chip is a 14-stage ripple binary counter, which has the added benefit of having its own internal clock generator. Because it is a CMOS device, it has a wide operating voltage, and in this circuit a 9V battery is being used to drive the circuit (although lower voltages could be used). The speed of the IC's internal oscillator is dependent on the values of three components R1, R2, and C1. R1 is also connected to a variable resistor, VR1, which will allow you to alter and fine-tune the speed of the internal oscillator even further.

The way that the circuit works is that once switch SW1 is activated, the 4060 starts to oscillate and produces a binary count on its output pins. This project uses the first three binary outputs from the chip, which are produced on pins 4, 5, and 7, to provide the various input combinations to the RGB LED. Unlike the 555 timer, the outputs of the 4060 are unable to continually source the amount of current that is required to drive the LED, so you therefore need to boost the current for each output. You do this by connecting each of the outputs to the base connections of three NPN transistors (TR1–TR3) via base resistors R3 to R5. Activating the base of the transistor basically switches the positive voltage rail through its collector emitter junction directly to each of the three anodes of the LED. This means that only a very small current is sourced from each of the ICs outputs, and these are used to switch a larger current to each of the LED colors.

Table 6-1 shows the various binary output combinations produced by the 4060 IC, and the different color combinations that the color-changing light box then produces. A value of 1 in the table denotes that the anode is activated, and a value of 0 means that the anode is switched off. Which colors are actually produced by the LED

TABLE 6-1 Color Combinations Produced by the RGB LED

Green (Output Q6)	Red (Output Q5)	Blue (Output Q4)	LED Color Output	Approx. Current Draw for Whole Circuit
0	0	0	None; the LED is off	0.5mA
0	0	1	Blue	14.5mA
0	1	0	Red	14.5mA
0	1	1	Purple	27mA
1	0	0	Green	14.5mA
1	0	1	Cyan	27mA
1	1	0	Orange	27mA
1	1	1	White/Blue White	38.5mA

depends on the intensity of each color element, and these are determined by the values of each of the LED series resistors that are used (R6–R8). Once the circuit is activated, the 4060 IC continuously cycles through each of the binary outputs and then starts at the beginning again.

Diode D2 is a rectifier diode and has been included in the circuit to prevent the components from being damaged in the event that the battery is accidentally connected the wrong way around.

Parts List

NOTE The Supplier and Part Number column of the following table lists specific parts that I used in this project. Refer to the appendix for additional details about acquiring your parts.

The parts you'll need for the color-changing light box project are listed in the following table.

PARTS LIST			
Code	Quantity	Description	Supplier and Part Number
IC1	1	4060B 14-bit binary counter	RS Components 308-938 (or similar)
D1	1	5mm RGB LED Red: V_F (typical) = 2.0V, I_F (max) = 30mA Green: V_F (typical) = 2.2V, I_F (max) = 25mA Blue: V_F (typical) = 4V, I_F (max) = 30mA	RS Components 247-1511
D2	1	1N4004 diode	—
R1	1	10KΩ 0.5W ±5% tolerance carbon film resistor	—
R2	1	1MΩ 0.5W ±5% tolerance carbon film resistor	—
R3–R5	3	1KΩ 0.5W ±5% tolerance carbon film resistor	—
R6*	1	270Ω 0.5W ±5% tolerance carbon film resistor	—
R7, R8*	2	390Ω 0.5W ±5% tolerance carbon film resistor	—

(continued)

PARTS LIST *(continued)*			
Code	Quantity	Description	Supplier and Part Number
VR1	1	1MΩ miniature enclosed horizontal preset potentiometer (min. 0.15W rated)	—
TR1–TR3	3	BC108C NPN transistor	—
C1	1	1μF 63V boxed polyester capacitor	—
SW1	1	Single-pole panel mount toggle switch, 2A rated	RS Components 710-9674
Hardware	1	Stripboard, 0.1″ (2.54mm) hole pitch, 37 holes wide by 24 tracks high	—
Hardware	1	16-pin DIL socket	—
Hardware	1	PCB mount holder for 9V PP3 cell	RS Components 489-611
Hardware	1	9V PP3 battery	—
Hardware	1	Clear LED lens mount (optional)	RS Components 223-1593 (pack of 5)
Hardware	—	Enclosure, cable, cable tie pad and cable tie, M3 nylon screw and nut; see text	—

*Note: If you use an LED that has different V_F and I_F values to the one specified in the parts list, then you may need to alter the resistance and wattage values of these LED series resistors. Please refer to Chapter 2, which explains how to do this.

Stripboard Layout

The stripboard layout for the project is shown in Figure 6-7. You need to make 17 track cuts, including those that sit underneath the IC's DIL socket. The track cuts are shown as white rectangular blocks in Figure 6-7, and the six larger squares in the middle of the board show the solder points for the RGB LED (D1).

How to Build and Test the Board

NOTE Please refer to Chapter 1 for soldering tips and techniques and for generic stripboard building guidelines.

Build the stripboard layout based on Figure 6-7. You should end up with a stripboard that looks similar to the one shown in Figure 6-8. You will notice that I used a PCB-mounted PP3 battery holder and soldered it into place, which allows the battery to be fitted onto the board to save space. If you decide to use this method, then you will also need to create a hole in the stripboard underneath the holder to secure it firmly to the stripboard. I used a nylon M3 screw and nut to do this. Also solder two flying leads to the board and solder them to the normally open connections of the power switch SW1.

Figure 6-9 shows how I mounted the RGB LED. I made sure to leave the leads quite long so that the LED is closer to the center of the enclosure that I used, which helps to give a better light effect in the dark. If you use an RGB LED that is different from the one that I used, then you will need to make sure that the pin configurations of the one that you use are the same. If they are not, then you may need to make some adjustments to the layout, or mount the LED off the board using flying leads.

Before you insert the IC into the DIL socket, you can power up the board and test that the transistors work and that they activate each of the

Chapter 6 ■ Exploring Multicolor LEDs: Color-Changing Light Box 67

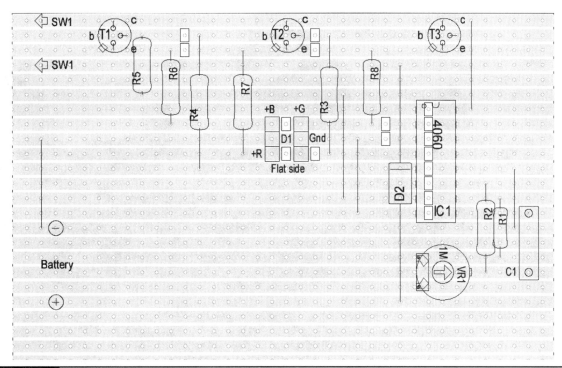

Figure 6-7 Stripboard layout for the color-changing light box

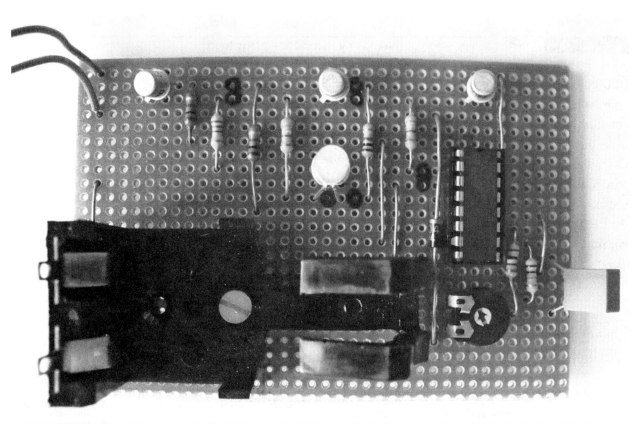

Figure 6-8 The finished stripboard. You will notice that D2 is wired before SW1 in my final prototype.

Figure 6-9 Elevate the position of the LED by leaving the leads long.

LED colors. First of all, connect the + lead of your multimeter to pin 16 of the DIL socket and connect the − lead to pin 8, to make sure that these pins are at 9V when the battery is fitted. If this is the case, you can then use a short piece of wire to short out pin 16 and pin 4 of the DIL socket, and this should illuminate the green part of the LED. Shorting out pins 16 and 5 should generate a red color, and shorting out pins 16 and 7 should create a blue color. Figure 6-10 shows how to do this.

Once you are happy with the results of the tests, remove the battery from its holder and insert the 4060 IC into the DIL socket (making sure that it is in the correct orientation) and turn the variable resistor fully counterclockwise. Now insert the battery again and turn on the switch; you should immediately see that the LED cycles through the seven color combinations at a fast rate. If this does not happen, remove the battery and perform the usual checks. If the circuit works, then you can slowly turn the variable resistor clockwise to slow down the rate at which the color changes. If you wind the resistor fully clockwise, you should find that it takes around 15 seconds for each color to change.

Finding an Enclosure

You now need to find a suitable enclosure for your light box. Ideally, you need to find an enclosure that is clear but has a sandblasted or milky look to it such that you cannot see what is in the box but the light from the LED can illuminate through it. I hunted around my garage and found an old box that used to hold business cards; it is semitransparent and ideal for this project. Figure 6-11 shows how I

Chapter 6 ■ Exploring Multicolor LEDs: Color-Changing Light Box

Figure 6-10 Making sure that the transistors and LED work

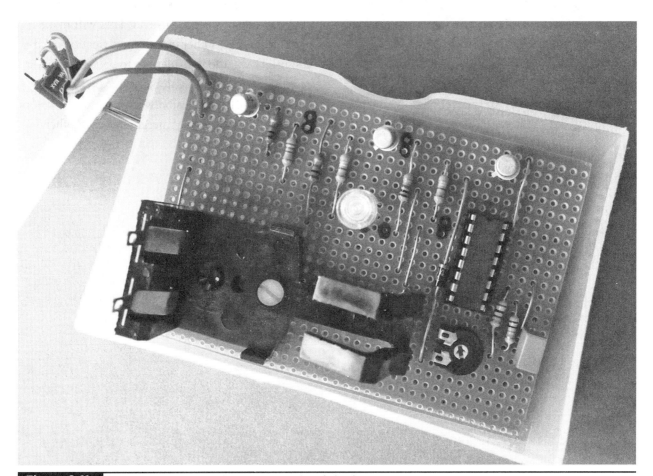

Figure 6-11 Mounting the switch onto the side of the box

mounted the power switch to the side of the box and secured the two flying leads using a cable tie base. I also mounted the stripboard in the lid of the box.

Figure 6-12 shows the completed project mounted inside the enclosure. I also decided to fit a clear lens onto the RGB LED to help diffuse the color output (you can see this in Figure 6-11), but this is not essential.

Figure 6-12 The finished light box

Light Me Up

You will only get the full effect of this project in a dark environment, so once you have built the unit, place it in a dark room and switch it on. The bright colors generated by the RGB LED diffuse through the milky enclosure to give the whole box a colorful glow, as shown in Figure 6-1 at the beginning of the chapter. Set the color cycle speed to a slow setting to provide a relaxing, color-changing mood light, or increase the speed to create a mini-disco light effect. Sit back, relax, and enjoy the light show!

Possible Circuit Modifications

If you decide that you want to alter the speed of the color changes on a regular basis, then you might prefer to remove the variable resistor from the stripboard and mount a panel-mounted version onto the side of the box instead. This will make it more accessible. Other circuit modifications that you might want to consider are to try and create an even bigger palette of colors, or maybe to soften the way the LED cycles between each of the colors so that the colors slowly merge into each other.

CHAPTER 7

Using Seven-Segment Displays: Mini Digital Display Scoreboard

THE TYPE OF LED THAT YOU ARE GOING TO EXPLORE in this chapter is the seven-segment display. This device allows you to display digital versions of the numbers 0 to 9 using LEDs. You likely have encountered these devices on many occasions. For example, the display on your DVD player or stereo system may use a version of an LED display, and if you have an LED alarm clock, it contains seven-segment displays. This type of LED device is normally a rectangular package that contains seven individual LEDs (eight if you include a decimal point), and the internal LEDs are positioned in such a way that you can create any number from 0 to 9, depending on which of the seven LEDs are illuminated. This chapter first introduces some of the seven-segment display variants that are available, and then shows you how to build a mini digital display scoreboard similar to the one shown in Figure 7-1.

Seven-Segment Displays

It will probably come as no surprise to you that seven-segment displays come in various shapes, sizes, and colors. The easiest type of seven-segment display to use is the single-digit version, but other variants are available that incorporate two or three seven-segment digits in a single package. Figure 7-2 shows a few examples of the LED displays that are available.

Each LED in a standard seven-segment display is identified by a letter (A to G), and the positions of LEDs A through G are fairly standard when looking at the various manufacturers' datasheets. The layout of a typical seven-segment display with its identifying letters is shown in Figure 7-3.

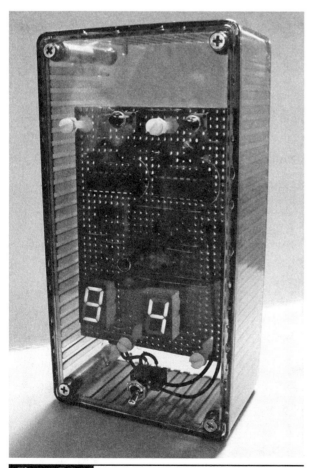

Figure 7-1 Mini digital display scoreboard project

71

Figure 7-2 LED seven-segment displays come in various shapes, sizes, and colors.

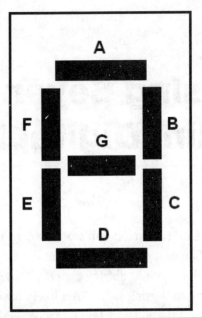

Figure 7-3 LED format for a seven-segment display

Which of the LED segments are illuminated determines which number, 0 to 9, is produced. Savvy hobbyists will also realize that this sort of device allows you to create letters as well as numbers, albeit a limited character set (for example, it is not easy to display the letter Q, W, or Z on a seven-segment display). In fact, there are other numerical LED displays that have more than seven segments, allowing you to create other characters and letters. An example of such a display is included in the top row of Figure 7-2. This type of display isn't covered in this book.

One of the main considerations when choosing a seven-segment device is which of the two main connection types to choose, as they are available in either common cathode configuration or common anode configuration. The LEDs inside a common cathode display have all seven cathodes linked together, and the LEDs in a common anode device have all seven anodes linked together (eight if you include the decimal point).

While referencing the LED layout in Figure 7-3, read through Table 7-1, which shows how you can generate the numbers 0 to 9 by activating various combinations of the seven LEDs (A–G). A value of

TABLE 7-1	Seven-Segment Display Code						
Displayed Number	A	B	C	D	E	F	G
0	1	1	1	1	1	1	0
1	0	1	1	0	0	0	0
2	1	1	0	1	1	0	1
3	1	1	1	1	0	0	1
4	0	1	1	0	0	1	1
5	1	0	1	1	0	1	1
6	1	0	1	1	1	1	1
7	1	1	1	0	0	0	0
8	1	1	1	1	1	1	1
9	1	1	1	1	0	1	1

1 in the table denotes that the LED is illuminated, and a value of 0 means that the LED is off.

Table 7-1 shows that a 7-bit binary code will allow you to produce numerical values 0 to 9 on a seven-segment display, so all you need now is a device that will generate a suitable 7-bit binary count like the one in Table 7-1. Fortunately, there are integrated circuits (ICs) available that make this job quite easy. Two examples are the 4026B and 4511B ICs, which are both CMOS devices. These will be discussed in the upcoming project. Another method that you can use to drive seven-segment displays is to use a microcontroller and program it with the required code to produce numbers on the display. You could also consider using multiplexing techniques to illuminate multiple displays. (Sorry, I got a bit carried away there and jumped ahead of myself! You'll come across this technique in later chapters.)

Now that you understand the concepts of illuminating a seven-segment display, you are ready to make your seven-segment display scoreboard.

Project 6
Mini Digital Display Scoreboard

The mini digital display scoreboard that you'll build in this project generates numerical values on two seven-segment displays and is designed to have two modes of operation.

How the Circuit Works

As mentioned earlier, some ICs are designed purposely to drive seven-segment displays. I initially considered the 4511B IC because it has good current-sourcing capabilities to drive LEDs brightly. However, it also requires a 4-bit binary coded decimal (BCD) counter to make it count

> **PROJECT SPECIFICATIONS**
>
> - The mini digital display scoreboard uses two separate seven-segment LED displays.
> - Two separate count buttons allow you to increment each of the displays.
> - A third button allows you to reset the display to 0.
> - In mode 1, each display can be individually made to count from 0 to 9. This could be used, for example, to keep score of a two-player game such as table soccer.
> - In mode 2, the two displays are linked and the display can be made to count from 0 to 99. (Alternatively, two separate scoreboards could be built, one per player.)

through the numbers 0 to 9. I wanted to keep the chip count as low as possible, so I decided to use the 4026B device instead. The 4026B device is ideal for this project because the single IC produces the required 7-bit binary code to create numbers on the seven-segment display, and it can also be "clocked" without the need for an additional counter IC. The only downside to using the 4026B is that its current-sourcing capabilities are not as good as the 4511B; in fact, each output can drive only a few milliamps depending on the supply voltage. If you wanted to drive the LED display brightly, you would need to incorporate some additional drive circuitry, such as a bank of transistors. However, because of the low current properties of LEDs, a usable light output can be produced even when the LEDs are supplied with limited current, so the 4026B device will work fine in this project.

NOTE *The 4026B IC only counts upward, so if you want to experiment with the circuit to build a digital counter that is able to count up or down, you'll need to use a different IC.*

Figure 7-4 shows the final circuit diagram for the scoreboard.

The circuit operation is fairly straightforward because most of the work is done by the two 4026B ICs. Both ICs connect directly to the seven-segment displays via series resistors R1 to R14. These resistors are connected to the seven output signals generated from each 4026B IC (pins 6, 7, 9, 10, 11, 12, and 13), which provide a high output signal to drive each of the individual LEDs in the seven-segment displays, and as you

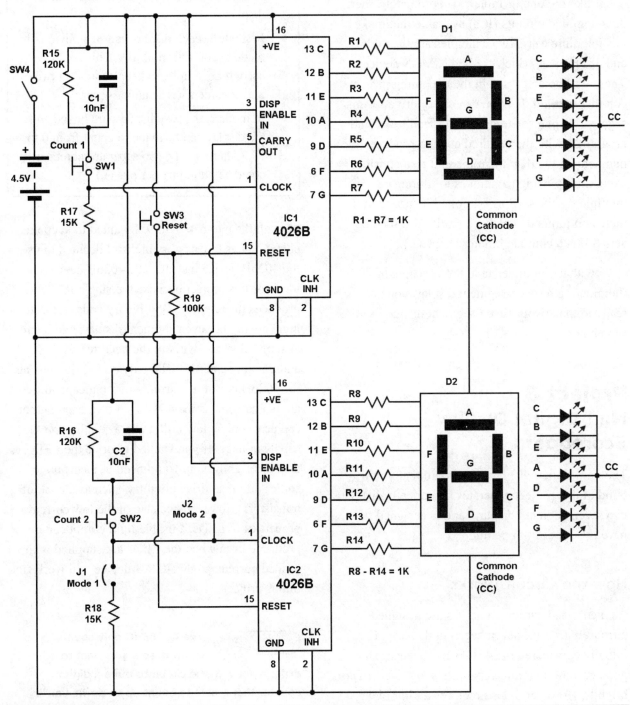

Figure 7-4 Circuit diagram for the seven-segment display scoreboard

can see the displays that are required are common cathode types. You can also see that both display circuits are identical and that they both share reset signals and power lines. Assuming that jumper link 1 is in place when the circuit is activated (mode 1), when push button switch SW1 is pressed, the number on display D1 increments by one count to show a number 1; pressing it again produces a number 2, and so forth. Pressing push button switch SW2 produces the same type of count on display D2. If either of the displays shows the number 9, it will roll back to 0 on the next press of the associated push button. This mode of operation creates a mini scoreboard that allows each player to score up to 9 points.

If you remove jumper link 1 and put it across jumper link 2 to activate mode 2, this makes the circuit operate in a different way because it links the carry-out signal on pin 5 from IC1 to the clock input pin 1 of IC2. This means that each time switch SW1 is pressed, D1 counts upward, and when it rolls from 9 to 0, a carry-out pulse from IC1 clocks IC2 to create a number 1 on D2. This gives the effect of a number 10 across both displays. This mode therefore allows you to create the numbers 0 to 99 using the two displays.

In either mode, both displays can be reset by pressing the reset button, SW3. Normally, both reset pins 15 are held low via resistor R19, but when the reset button is pressed, both IC reset pins are taken high, which resets both displays to 0. The clock inhibit pin 2 on both ICs is held low to allow counting to take place.

You will also notice that there are two resistors and a capacitor around each of the clock and switch circuits, these being R15, C1, and R17 for IC1 and R16, C2, and R18 for IC2. These are included in the circuit to debounce the count pulse from the switches when you press them. Without these components, the display could increment by two or three digits whenever you press the count buttons, because the contacts of a mechanical switch can "bounce" a number of times when you press it, and this can be seen as multiple pulses by a digital clock circuit. If you find that your switch still counts erratically, you could try increasing the capacitor values of C1 and C2 to see if this fixes the problem. Notice that when you remove jumper link 1 and place it on jumper link 2 to generate mode 2, the clock input to IC2 is no longer held low circuit via R18; this allows the carry-out signal from IC1 to clock IC2 in the required manner.

As mentioned earlier, the display outputs from the 4026B can drive only a small amount of current, and this is why the series resistors for each LED are rated at 1KΩ. Although this reduces the current supplied to each LED segment to less than 2.5mA, the LED display still illuminates sufficiently for you to read the display. The whole circuit is driven by a 4.5V supply, which is produced by three AA batteries.

Parts List

NOTE The Supplier and Part Number column of the following table lists specific parts that I used in this project. Refer to the appendix for additional details about acquiring your parts.

The parts you'll need for the mini digital display scoreboard project are listed in the table that follows.

PARTS LIST

Code	Quantity	Description	Supplier and Part Number
IC1, 2	2	4026B decade counter with seven-segment display outputs	ESR Electronic Components 4026B
D1, D2	2	Seven-segment green common cathode display V_F (typical) = 2.1V, I_F (max) = 30mA	RS Components 195-108 (Mfr: Avago Tech. HDSP-5603)
R1–R14*	14	1KΩ 0.5W ±5% tolerance carbon film resistor	—
R15, R16	2	120KΩ 0.5W ±5% tolerance carbon film resistor	—
R17, R18	2	15KΩ 0.5W ±5% tolerance carbon film resistor	—
R19	1	100KΩ 0.5W ±5% tolerance carbon film resistor	—
C1, C2	2	10nF ceramic disk capacitor (min. 16V rated)	—
SW1–SW3	3	0.24″ × 0.24″ (6mm × 6mm) tactile momentary push-to-make switch 0.67″ high (17mm), 50mA rated	RS Components 479-1463 (pack of 20)
SW4	1	Single-pole panel mount toggle switch, 2A rated	RS Components 710-9674
Hardware	1	Stripboard, 0.1″ (2.54mm) hole pitch, 37 holes wide by 24 tracks high	—
Hardware	2	16-pin DIL socket	—
Hardware (J1 and J2)	1	Four-way, single-row PCB pin header	ESR Electronic Components 111-104
Hardware (J1 and J2)	1	Jumper link to suit single-row PCB pin header	ESR Electronic Components 111-930
Hardware	1	20-way turned pin SIL socket strip	RS Components 267-7400 (pack of 5)
Hardware	1	Enclosure: translucent blue 5.9″ (150mm) × 3.15″ (80mm) × 2″ (50mm)	RS Components 528-6906 (Mfr: Hammond 1591DTBU)
Hardware	1	PP3 battery clip and lead	RS Components 489-021 (pack of 5)
Hardware	1	AA battery holder (three AA batteries)	Maplin YR61R
Hardware	3	AA battery (1.5V)	—
Hardware	4	M3 × 20mm nylon slot screws	RS Components 527-656 (bag of 100)
Hardware	12	M3 nylon nuts	RS Components 525-701 (bag of 50)
Hardware	—	Double-sided self-adhesive tape, cable tie	—

* Note: If you use seven-segment displays that have different V_F and I_F values to those that are in the parts list, then you may need to alter the resistance and wattage values of these LED series resistors. Please refer to Chapter 2, which explains how to do this, and also remember that you should be using a value of less than 2.5mA (ideally 1mA if possible) for the value of I_F in your calculations for this project.

Stripboard Layout

> **NOTE** Please refer to Chapter 1 for soldering tips and techniques and for generic stripboard building guidelines.

The stripboard layout for the mini digital display scoreboard is shown in Figure 7-5. If you've been tackling the projects in this book in order, this is the busiest stripboard layout that you have encountered so far. The reason this is so crowded is that all of the components are mounted on a single piece of stripboard. You might decide to mount the two LED displays on a separate piece of stripboard, but then you would need to solder interconnecting wires to each of the LED segments, possibly by using a length of ribbon cable.

You might also decide to use a different type of seven-segment display with different pin-outs from the type that I used. In that case, you will need to make some modifications to the stripboard layout, and possibly to the values of the series resistors, to suit the device that you decide to use.

Note that you need to make 50 track cuts on the stripboard, including those that sit beneath the IC sockets and the seven-segment displays; these are shown as white rectangular blocks in Figure 7-5. You also need to drill four 3mm holes in the stripboard, identified by the four white circles with a + sign.

How to Build and Test the Board

Carefully follow the stripboard layout shown in Figure 7-5 to build the circuit. You create J1 and J2 by cutting the four-way pin header in half to create two two-way pin headers, which you then need to

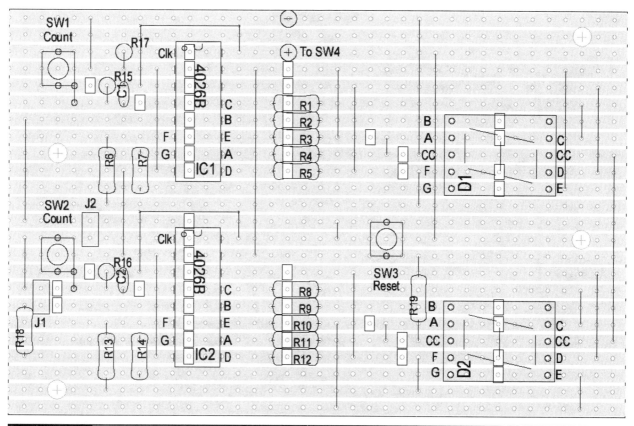

Figure 7-5 Stripboard layout for the mini digital display scoreboard

solder in place. I also decided to use turned pin SIL sockets for the seven-segment displays rather than soldering the displays directly to the board. Cut a 20-way SIL socket into four 5-way SIL socket strips and solder these into place. You can the push the pins of the seven-segment displays onto the SIL sockets once the board is complete. Your stripboard should end up looking like the one shown in Figures 7-6 and 7-7.

Figure 7-7 shows how I positioned the 14 resistors R1 to R14 in an alternating fashion; this was to help avoid the resistor leads shorting out on each other if they were accidentally pushed together. Also note that I had to fit the wire links underneath the two switches, SW1 and SW2, before I soldered the switches in place (these can be seen on the left side of both Figures 7-6 and 7-7).

Before you insert the two 4026B ICs, you can test the board in a similar manner to the way you tested the color-changing light box project in Chapter 6 (if you completed that project). When you apply the 4.5V supply to the board without the ICs inserted, none of the displays should illuminate. Using a multimeter, measure the voltage across pins 16 and 8 of both IC sockets; you should find that pin 16 of both ICs is + 4.5V with respect to pin 8. You can also check that each segment of the displays works okay. For example, shorting out pin 16 (+) to pin 13 of IC1's DIL socket should illuminate segment C of display D1, and shorting out pin 16 (+) to pin 12 of IC1's DIL socket should illuminate segment B of display D1. You can follow this method to prove that all 14 segments across both displays operate in the correct manner.

Figure 7-6 The completed stripboard layout

Chapter 7 ■ Using Seven-Segment Displays: Mini Digital Display Scoreboard

Figure 7-7 Note the resistor positions and the wire links under the switches.

Once you are happy with the results of the tests, remove the battery from the circuit and insert the two 4026B ICs into their DIL sockets, making sure that they are inserted in the correct orientation. Then insert the jumper link across the J1 pin header and connect the battery to the circuit; you should find that both displays show the number 0. Pressing SW1 once should increment the count on D1 by one, and pressing SW2 once should do the same for D2. Keep pressing SW1 to make sure that display D1 shows all of the numbers 0 to 9 correctly, and then do the same for display D2 by pressing SW2. If your circuit does not operate in this manner, then you need to check your stripboard over to make sure that the wire links and components are soldered in the correct place and that none of the tracks are linked by stray solder. If you find that a single switch press causes the count to increment by 2 or 3 counts, then you may need to adjust the values of the resistor capacitor debounce network to suit your switch type. Sometimes, pressing the switch gently also helps to avoid debounce, but it all depends on the type of switch that you use.

If everything works okay, press the reset switch SW3, and both displays should reset to 0. Now remove the jumper link from J1 and insert it across the J2 pin header instead. You should find that if you keep pressing switch SW1, you can make the two displays work as one to count from 0 to 99 and back to 0 again.

Mounting the Board in an Enclosure

I decided to use a blue, transparent enclosure for this project, which allows me to see the electronics inside the box, helps to act as a filter for the two green displays, and seems to make the display look clearer. If you decide to use a clear box like this one, I recommend marking out the position of the drill points on the lid for the four switches and the four fixing screws. Figure 7-8 shows a close-up of the lid, into which I have drilled holes and have inserted the four M3 nylon screws into the fixing holes. You can also see on the right side of the figure that I have fitted the power switch (SW4) into place.

I used the four nylon screws and nuts to create stand-offs to mount the stripboard to the lid. The four nuts need to be positioned low enough for the board-mounted switches to protrude through the holes in the lid. The board then fits onto the four nylon screws and rests onto the four nylon nuts. I then used four additional nylon nuts to secure the stripboard into place, as shown in Figure 7-9.

CAUTION Make sure that you use insulated screws and nuts to mount the board to the enclosure. Stripboard tracks will be shorted out if you use metal screws and nuts, which will cause the circuit to malfunction and may even damage the components.

Now cut the positive supply cable between the battery connector and the board, and then solder these two positive flying cables to the power

Figure 7-8 The drilled lid, complete with nylon screws and nuts

Chapter 7 ■ Using Seven-Segment Displays: Mini Digital Display Scoreboard 81

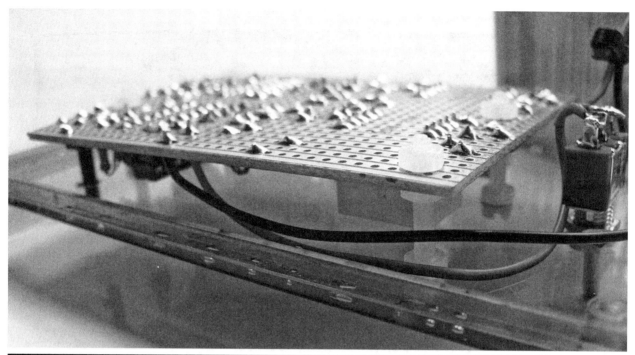

Figure 7-9 The stripboard is mounted onto the nylon screws and fixed in place.

switch, SW4. Secure the AA battery enclosure to the inside of the box by using double-sided adhesive tape, and then insert the batteries. You should end up with a finished enclosure and lid similar to those shown in Figure 7-10.

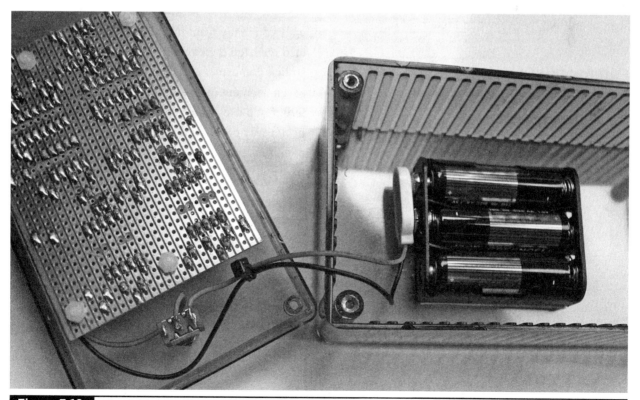

Figure 7-10 The lid is complete and the battery holder is secured to the box.

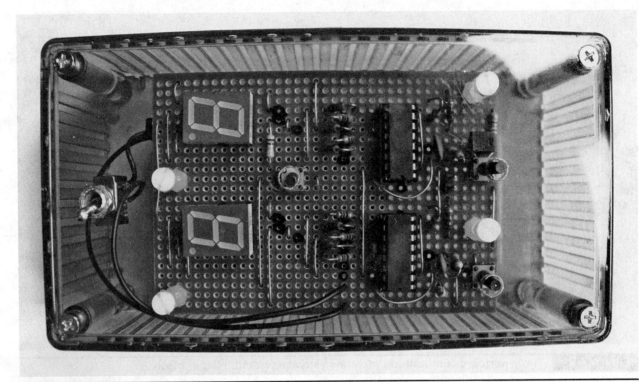

Figure 7-11 The final scoreboard

Finally, screw the lid to the enclosure using the screws provided, and you will have a finished scoreboard like that shown in Figure 7-11.

Future Modifications

For convenience, you might decide to replace the board-mounted jumper pins J1 and J2 with a single "mode" switch mounted onto the front of the enclosure. That way, you won't have to remove the stripboard from the lid every time you want to change the mode of operation.

If you really want to be adventurous, you can also redesign the circuit and make a larger board so that you can expand the circuit to drive three seven-segment displays, allowing a count from 0 to 999 if you use a third 4026B IC (there is a much neater way of doing this, which we shall explore in a later chapter).

PART TWO
Sequencer Projects

CHAPTER 8

Introducing the 4017 Decade Counter: Experimental LED Sequencer Circuit

THE PROJECTS IN THE FIRST PART OF THIS BOOK involved basic illumination and flashing of LEDs. The project presented in this chapter is the first in a series of projects that incorporate a sequencer circuit. That is, these projects contain several LEDs that illuminate in sequence, either one or more at a time. These sequencer circuits could also be called "cascade" or "chaser" circuits, because sometimes the visual effect created appears as though the string of lights cascades or the individual lights chase each other in a loop. If you have ever been to a fair, a theatre, or the Las Vegas Strip, you probably have seen sequencer circuits in action on a marquee advertising a show, where the signage is surrounded by a row of colored lights that either flash or seem to chase each other in sequence.

> **NOTE** If you completed the project in Chapter 6, then you've already encountered a circuit that uses a basic sequencer circuit to illuminate LEDs, although this may not have been immediately obvious. The color-changing light box project uses a 4060 IC to output a binary counting sequence, and three of its binary outputs are used to change the color of an RGB LED.

This chapter starts by showing you how to build an experimental sequencer circuit like the one shown in Figure 8-1. It then explores some of the things that you can do with the circuit. The board could also be used as the basis of a larger LED flashing display. The subsequent chapters in this part of the book show you how to build some more-exciting sequencer projects.

The 74HC Range of ICs

The 4000 series IC devices, such as the 4017 which is discussed shortly, and those that were used in the projects in Chapter 6 (4060) and Chapter 7 (4026B), have limited output current-sourcing capabilities, requiring that you either provide additional driver circuitry to boost the current or increase the series resistor value to reduce the LED current drawn. One benefit of using the CMOS 4000 range of ICs is that they will accept a good range of supply voltages, typically 3–18 volts, but that is compromised by their current output capabilities, which are limited to only a few milliamps.

Another range of 4000 series ICs that is available is the 74HC type. One benefit of using the 74HC range ICs is that their output current capabilities are sometimes higher than those of the 4000 version. The drawback is that the maximum

Figure 8-1 Experimental sequencer circuit board

supply voltage allowed with these devices is usually 5V or 6V. You will find that some, but not all, of the 4000 CMOS devices are duplicated in the 74HC range. Be aware, though, that the pin-outs of equivalent devices do sometimes vary between the two ranges, and it is always advisable to check the manufacturer's datasheet to confirm the pin-outs instead of assuming that they are pin compatible. Fortunately, the 4017 decade counter which we are going to use in this project is available in a 74HC version, and its output pins are capable of sourcing up to 25mA, which means that you do not need to incorporate any additional drive circuitry to operate a single LED per output. In theory, it is possible to use a standard 4017 IC in this project because the pin-outs of both of these devices are the same, but to do so, you would need to make sure that you increase the value of the LED series resistor R4 to ensure that the LED current is reduced to only 1 or 2 milliamps; otherwise, you will damage the IC.

Project 7
Experimental LED Sequencer Circuit

The project in this chapter introduces you to an integrated circuit that, in my opinion, is one of the most versatile devices that can be used at the heart of a sequencer circuit, this being the 4017 decade counter.

PROJECT SPECIFICATIONS
■ The experimental board uses an LED bar-graph display to demonstrate the operation of a sequencer circuit.
■ The operation of the circuit can be easily modified without having to desolder any of the components from the board.
■ The experimental LED sequencer circuit could be used at the heart of a much bigger LED project.

The IC used in this project is the 74HC variant of the 4017 decade counter, which contains ten individual decoded outputs; this means that there is only ever one output with a high state at any one time. Every time the 4017 receives a clock signal from an external clock circuit, the next output in the sequence goes high. You can therefore use these ten decoded outputs to control individual LEDs and produce some interesting illumination effects, as you will see. As mentioned, the 4017 device does require a separate clock circuit to trigger the next output, unlike the 4060 device (used in the color-changing light box project), which has its own internal oscillator. Table 8-1 shows how the ten outputs (Q0 to Q9) of the 4017 IC operate whenever the IC is clocked. A value of 1 in the table denotes that the output is high, and a value of 0 means that it is low.

How the Circuit Works

Figure 8-2 shows the circuit diagram for the experimental LED sequencer circuit.

You will be using two ICs in this project: IC1 is the clock circuit and IC2 is the sequencer. As in previous projects, the circuitry around IC1 is a 555 timer in astable mode, and its output clock speed is determined by the values of R1, R2, VR1, and C1. The clock output of IC1 is pin 3, and this is connected directly to the clock input pin 14 of IC2, which is a 74HC4017 decade counter. Whenever pin 14 of IC2 receives a clock signal from the 555 timer, it activates one of the ten outputs listed in Table 8-1. The reset pin of the 74HC4017 is kept low via resistor R3, and if this pin is made to go high, the decade counter resets to its starting output, Q0 (pin 3). Pin 13 of IC2 is the "clock enable" pin, and this needs to be kept low for the counter to operate. The only pin that is not used in this circuit is pin 12, which is a "counter out" pin; it can be used to cascade multiple 74HC4017 ICs together, allowing you to illuminate a sequence of more than ten LEDs if required.

The ten decoded outputs from IC2 are connected to a 20-pin DIL socket that will allow you to fit D1, which is a ten-segment bar graph LED. This project is the first in which this type of

TABLE 8-1	How the Outputs of the 4017 IC Operate									
Clock Pulse	Q0	Q1	Q2	Q3	Q4	Q5	Q6	Q7	Q8	Q9
1	1	0	0	0	0	0	0	0	0	0
2	0	1	0	0	0	0	0	0	0	0
3	0	0	1	0	0	0	0	0	0	0
4	0	0	0	1	0	0	0	0	0	0
5	0	0	0	0	1	0	0	0	0	0
6	0	0	0	0	0	1	0	0	0	0
7	0	0	0	0	0	0	1	0	0	0
8	0	0	0	0	0	0	0	1	0	0
9	0	0	0	0	0	0	0	0	1	0
10	0	0	0	0	0	0	0	0	0	1
11	1	0	0	0	0	0	0	0	0	0
12	0	1	0	0	0	0	0	0	0	0

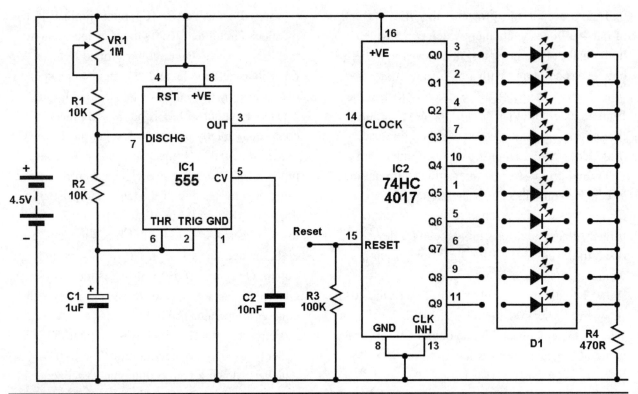

Figure 8-2 Circuit diagram for the experimental LED sequencer circuit

LED package is used. These devices normally contain 10 or 20 individual LEDs conveniently encapsulated into a single package. Normally, they are used in circuits that require a bar graph type of display, such as a graphic equalizer in a mixer desk or a stereo system. This project uses one with ten LEDs to demonstrate how the 74HC4017 IC illuminates a row of LEDs. Also notice that there is only one series resistor to limit the current to the ten LEDs, this being R4. The reason for this is that only one LED is illuminated at any one time, so a single resistor is sufficient.

Parts List

NOTE The Supplier and Part Number column of the following table lists specific parts that I used in this project. Refer to the appendix for additional details about acquiring your parts.

The parts you'll need for the experimental LED sequencer circuit project are listed in the upcoming table.

Stripboard Layout

The stripboard layout for the experimental LED sequencer circuit is shown in Figure 8-3, which shows the position of the components and the 29 track cuts that are required (these are shown as white rectangular blocks).

Notice at the bottom of the board that there is also a ten-way single in-line (SIL) socket positioned next to the 20-pin dual in-line (DIL) socket. The ten-segment LED display fits into the 20-way DIL socket, which means that the display can be removed and replaced easily if required. The ten-way turned pin SIL socket is used to connect interconnecting cables to the board to

PARTS LIST

Code	Quantity	Description	Supplier and Part Number
IC1	1	555 timer	RS Components 534-3469 (or similar)
IC2	1	74HC4017 decade counter	RS Components 709-3062 (pack of 10) or ESR Electronic Components 74HC4017
R1, R2	2	10KΩ 0.5W ±5% tolerance carbon film resistor	—
R3	1	100KΩ 0.5W ±5% tolerance carbon film resistor	—
R4*	1	470Ω 0.5W ±5% tolerance carbon film resistor	—
VR1	1	1MΩ miniature enclosed horizontal preset potentiometer (min. 0.15W rated)	—
C1	1	1μF 16V radial electrolytic capacitor	—
C2	1	10nF 16V ceramic disk capacitor	—
D1	1	Ten-segment LED green bar graph display (see text) V_F (typical) = 2.1V, I_F = 20mA (typical)	RS Components 719-2409 (pack of 2) (Mfr: Kingbright DC-10CGKWA)
D1 (alternative)	1	Seven-segment red common cathode display HDSP-5503 (see text) V_F (typical) = 2.1V, I_F (typical) = 20mA	RS Components 587-951 (Mfr: Avago Tech. HDSP-5503)
Hardware	1	Stripboard, 0.1" (2.54mm) hole pitch, 37 holes wide by 24 tracks high	—
Hardware	1	8-pin DIL socket	—
Hardware	1	16-pin DIL socket	—
Hardware	1	20-pin DIL socket (turned pin version)	—
Hardware	1	20-way turned pin SIL socket strip	RS Components 267-7400 (pack of 5)
Hardware	1	PP3 battery clip and lead	RS Components 489-021 (pack of 5)
Hardware	1	AA battery holder (three AA batteries)	Maplin YR61R
Hardware	3	AA battery (1.5V)	—
Hardware	—	Solid interconnecting wires	—
Hardware	1	Breadboard	—

*Note: If you use LED displays that have different V_F and I_F values to those specified in the parts list, then you may need to alter the resistance and wattage values of this LED series resistor. Please refer to Chapter 2, which explains how to do this.

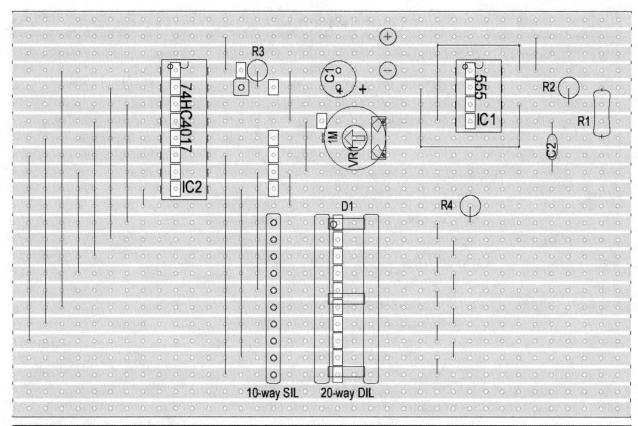

Figure 8-3 The experimental LED sequencer stripboard layout

experiment with the circuit, which we shall see later. There is also a single turned pin socket that is positioned next to the reset pin 15 of IC2. You can create the 10-pin and 1-pin SIL sockets required by cutting a 20-pin SIL socket strip down to size.

How to Build and Test the Board

NOTE Please refer to Chapter 1 for soldering tips and techniques and for generic stripboard building guidelines.

Build the sequencer circuit by following the stripboard layout shown in Figure 8-3. You should end up with a stripboard that looks like the stripboard shown in Figures 8-4 and 8-5.

Before you fit the two ICs into their sockets, insert the ten-segment LED display into the 20-pin DIL socket, making sure that the anode side (+) of the display is to the left of the board (use the LED display datasheet to help you identify the anode pins). You can now perform some tests to make sure that the LED section works okay. Figure 8-6 shows how to connect a piece of link wire between pin 16 of IC2's DIL socket and the top pin (pin 1) of the SIL socket. Before you do this, apply the 4.5V battery to the circuit and make sure that the voltages at the DIL sockets of pins 1 and 8 of IC1 and pins 16 and 8 of IC2 are in line with the supply voltage as expected. If these are okay, then use the wire link as shown, and this should illuminate the top LED of the segment display. Next, keep the link connected to pin 16 (+) of IC2's DIL socket and move the other end of the link down to pin 2 of the SIL socket. This should now illuminate the next LED down in the segment display.

Chapter 8 ■ Introducing the 4017 Decade Counter   91

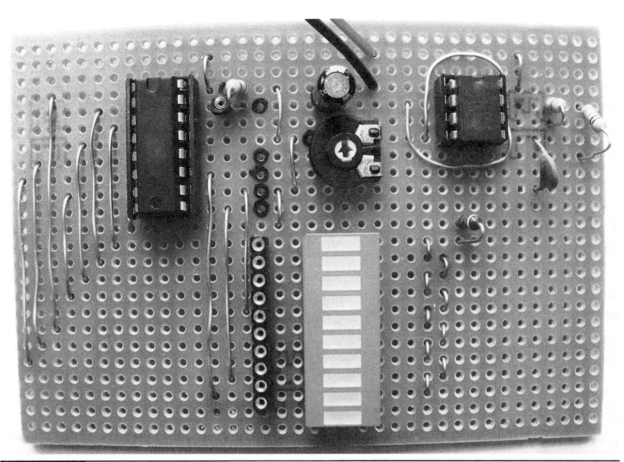

Figure 8-4   Completed stripboard layout

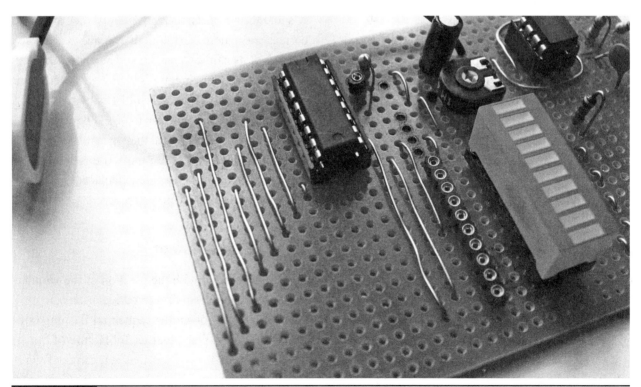

Figure 8-5   Close-up of the wiring around the 74HC4017 IC

Figure 8-6 Use a wire link to test each of the LEDs in turn.

Keep on moving the wire link down the SIL socket until you have tested all ten LEDs in the display. If multiple LEDs illuminate or no LEDs illuminate, then something is wrong, possibly solder bridges in between adjoining tracks.

The next step is to do the same tests but connect the link wire between pin 3 and pin 16 of IC2 instead. This should again illuminate the top LED of the display. You can now work through each of the output pins of IC2 in turn to make sure they illuminate the LED segments in sequence from Q0 to Q9. Always keeping one end of the wire connected to positive pin 16, test each of the pins 3, 2, 4, 7, 10, 1, 5, 6, 9, and 11 in turn. This emulates the 74HC4017 chip when it is being clocked by the 555 timer.

Once you are happy with the results of all the checks, remove the battery and then insert the two ICs into their sockets, making sure that they are in the correct orientation. Next, turn the variable resistor so that it is at its halfway position and reconnect the battery. If the timer circuitry is working correctly, you should find that each of the ten LEDs illuminates sequentially, starting from the top LED down to the bottom and then repeating at the top again (as is occurring in Figure 8-7; trust me!). If this doesn't happen, then you may have some faults around the 555 timer to resolve before you proceed.

Time to Experiment

If you play around with the position of the variable resistor by using a small screwdriver, you can speed up or slow down the sequential flashing rate of the ten LEDs. One other useful feature of the 74HC4017 and 4017 ICs is that you can configure them to count in sequence from any number

Figure 8-7 The LEDs illuminate one at a time in sequence.

between 2 and 10. At the moment, the circuit is configured to count to 10, but if you connect a wire link between the one-way SIL socket next to reset pin 15 of IC2 and pin 7 of the ten-way SIL socket (Q6 pin 5 of IC2), as shown in Figure 8-8, you can make the LEDs count to 6 instead.

What is happening here is that whenever the seventh pin of IC2 goes high, it immediately sends reset pin 15 high, which in turn resets the counter. Now connect the link wire from the reset pin 15 to pin 4 of the SIL socket, as shown in Figure 8-9. This converts the circuit to count to 3, which means that only three LEDs are in the sequence.

The preceding experiments show the versatility of the 4017 counter. You can use this same principle to experiment with a seven-segment display and illuminate it in an unusual way. Remove the battery from the board and carefully remove the LED display from its DIL socket; you may need to use a small screwdriver to gently pry it from the socket. Ease it out gently by alternating the pressure of the screwdriver to the display one side at a time so that you don't damage the display or any of its pins.

Now insert a common cathode seven-segment display into a small piece of breadboard and, using solid interconnecting wires, connect it to the pins

94 Brilliant LED Projects

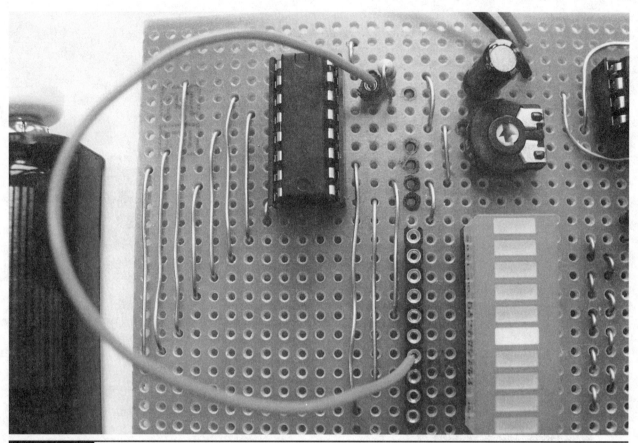

Figure 8-8 Converting the circuit to count to 6

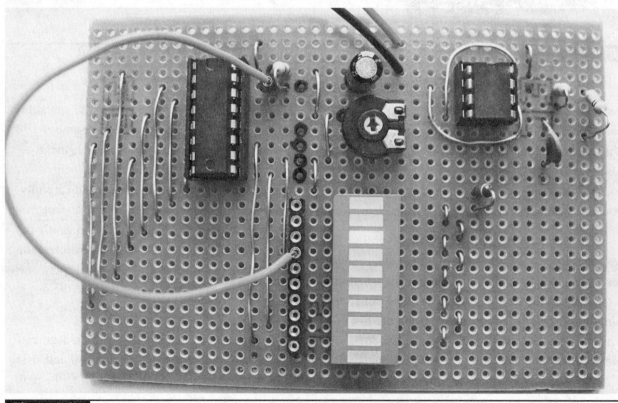

Figure 8-9 Counting to 3

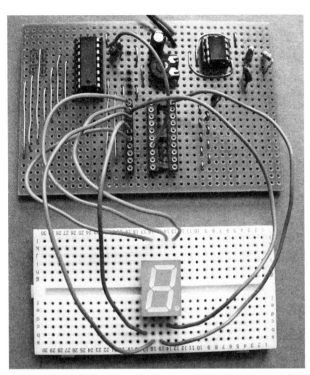

Figure 8-10 Connecting the seven-segment display to the experimental board

TABLE 8-2	Connections Between the Stripboard and the Seven-Segment Display
10- or 20-Way SIL Socket Number (1 is the top socket)	Seven-Segment Display LED Letter Connection
1	A
2	B
3	C
4	D
5	E
6	F
7	Connect to one-way SIL socket next to reset pin 15 of IC2 (74HC4017)

on the stripboard as shown in Figure 8-10. I used an HDSP-5503 device (RS part number 587-951). If you are using a seven-segment display that has different pin-outs from the one I used, then you will need to connect the interconnecting wires up differently.

The connections required between IC2 and the LED letters of the seven-segment display are shown in Table 8-2. Also note that you need to connect a wire between the common negative connection (pin 20 of D1's 20-way DIL socket) and the common cathode connection of the seven-segment display.

So, for example, by referring to Table 8-2, you can see that pin 1 of the ten-way SIL socket needs to be connected to the letter A connection of the seven-segment display.

Once you have connected the seven-segment display to the experimental sequencer board as outlined in Table 8-2, connect the battery to the board. What you have created here is a count-by-6 counter, and what you should see is that the sequencer circuit illuminates six of the seven segments in the seven-segment display in turn, creating the effect of a moving circle as each LED A, B, C, D, E, and F illuminates in turn. You can increase or decrease the speed of the display by adjusting the variable resistor VR1. Continue experimenting with the wiring configuration between pins 1 to 6 of the ten-way SIL socket and the seven-segment display to alter the shape of the flashing sequence of the display.

Further Modifications

The purpose of this chapter's project is to demonstrate a basic sequencer circuit, but you could use this board as the basis of a larger project. If you remove the ten-digit bar graph LED, you could solder interconnecting cables from the board and connect them to a string of ten external 5mm LEDs and produce a flashing light effect. The effect would look even flashier if you were to use various color LEDs in the string. You could even create a miniature festive light display. If you decide to do this, you will need to make sure that you alter the resistance and wattage value of R4 to accommodate each color LED.

As you probably realized, the source current of IC2 will allow you to drive only one bright LED per output comfortably. How could you get around this limitation if you wanted to drive a total of 24 LEDs from a single 4017 IC instead? The next project in the sequence shows you how!

CHAPTER 9

Driving Multiple LEDs from a Single IC Output: Color-Changing Disco Lights

THIS CHAPTER'S PROJECT IS AN EXAMPLE OF THE TYPE of LED circuit that you can create by modifying the experimental sequencer circuit built in project 7 in the preceding chapter. You can use the final product of this chapter's project to produce an eye-catching disco lighting effect, as shown in Figure 9-1, or you could even modify it to emulate the flashing control panel of a spaceship, just like those that you see in sci-fi films.

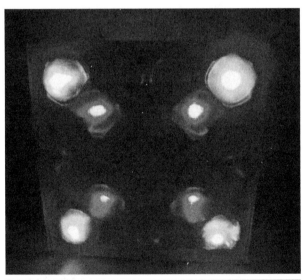

Figure 9-1 Color-changing disco light project

NOTE If you have not yet built project 7 it is recommended that you at least read Chapter 8 before you build this project, so you can understand the circuit concepts that have gone into creating this project.

Project 8
Color-Changing Disco Light

Project 7 in Chapter 8 demonstrated that the 74HC4017 decade counter IC is able to provide a total source current of 25mA from each of its outputs. That is sufficient if you want to drive a single LED brightly, but in this project you want to be able to drive up to four LEDs per counter output, which could equate to 70mA or 80mA per output, depending on the value of the series resistors that you use. Therefore, some additional drive circuitry is required for this project to boost the current outputs of the counter. The sequencer circuit will count to 7, so you could use seven separate transistors to boost the output current, but I'll show you an alternative method that reduces the overall component count of the final circuit.

> **PROJECT SPECIFICATIONS**
>
> - The result of the project is an eye-catching and colorful disco lighting effect.
> - The display includes a total of 24 colored LEDs, which flash in a unique sequence.
> - Five different LED colors are used in the display.
> - The project incorporates an adjustable-speed sequencer circuit using a count-to-7 configuration.
> - Additional drive circuitry allows the sequencer to drive up to four LEDs at a time.
> - The supply voltage is 4.5 volts DC.

How the Circuit Works

Figure 9-2 shows the circuit diagram for the color-changing disco light project. It contains many of the components that were used in project 7.

The circuitry around IC1 and IC2 is very similar to the circuitry in project 7, IC1 being the 555 timer that is used to clock IC2, a 74HC4017 decade counter, to create a sequential output. However, the circuit in this project has three main differences from the circuit in project 7.

The first difference is that pin 6 of IC2 is connected directly to reset pin 15, which means that IC2 counts to 7 rather than to 10, because the counter resets when the eighth counter output (Q7) goes high.

The second change is the inclusion of IC3, which is a ULN2003 integrated circuit; this chip is a Darlington transistor array that contains seven Darlington transistors in a single package. The way that the device works is that whenever a positive signal is applied to one of the base connections (for example, 1B), its corresponding collector output (in this case, 1C) goes low. Each of the seven base connections of the ULN2003 contains a 2.7KΩ resistor inside the device, so there is no need to provide additional base resistors in the circuit. Also, the collector outputs are able to drive 500mA, which is plenty to drive four LEDs at a time, the project requirement. The circuit diagram also shows that the seven outputs from IC3 are connected to the cathodes (−) of up to four LEDs at a time; you can configure these connections to make your own bespoke display layout. The 24 LEDs are wired up in a common anode (+) configuration via each LED's series resistor. The diagram does not show all 24 LEDs, but you will find out how to connect these up shortly.

Finally, notice that the diagram includes an additional electrolytic capacitor, C3, which is wired across the 4.5V battery supply. This capacitor is known as a *decoupling* capacitor, and it is recommended that you always include one in a circuit like this. Its purpose is to smooth the supply voltage. This helps to stop spurious triggering of the sensitive CMOS devices, which could occur if you don't include the decoupling capacitor. The circuit is drawing around 70mA from each output for a split second at every clock pulse, so the circuit does benefit from the inclusion of a decoupling capacitor. The value of C3 will depend on the total current drawn by the circuit, but in this case a 1000µF capacitor works well. One additional point to note is that because you are incorporating the LED driver (IC3), you could use a standard 4017 CMOS device for IC2 if you wanted to.

Parts List

> **NOTE** The Supplier and Part Number column of the following table lists specific parts that I used in this project. Refer to the appendix for additional details about acquiring your parts.

The parts you'll need for the color-changing disco light project are shown in the upcoming "Parts List" table.

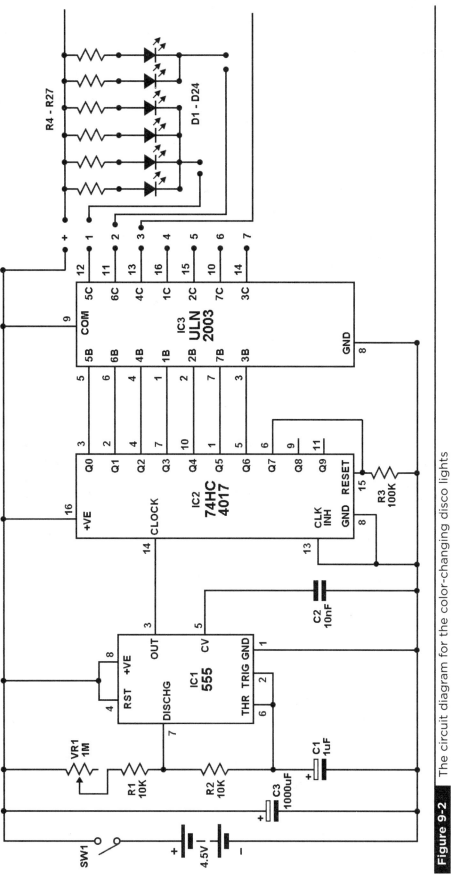

Figure 9-2 The circuit diagram for the color-changing disco lights

PARTS LIST

Code	Quantity	Description	Supplier and Part Number
IC1	1	555 timer	RS Components 534-3469 (or similar)
IC2	1	74HC4017 decade counter	RS Components 709-3062 (pack of 10) or ESR Electronic Components 74HC4017
IC3	1	ULN2003A Darlington transistor array	RS Components 436-8451
R1, R2	2	10KΩ 0.5W ±5% tolerance carbon film resistor	—
R3	1	100KΩ 0.5W ±5% tolerance carbon film resistor	—
R4–R27*	24	150Ω and 180Ω 0.5W ±5% tolerance carbon film resistor (see text)	—
VR1	1	1MΩ miniature enclosed horizontal preset potentiometer (min. 0.15W rated)	—
C1	1	1µF 10V radial electrolytic capacitor	—
C2	1	10nF ceramic disk capacitor (min. 16V rated)	—
C3	1	1000µF 10V radial electrolytic capacitor	—
D1–D24	24 total: 4 blue 5 red 5 orange 5 yellow 5 green	5mm LEDs, various colors: Blue: V_F (typical) = 4.0V, I_F (max) = 30mA Red: V_F (typical) = 2.0V, I_F (max) = 30mA Orange: V_F (typical) = 2.0V, I_F (max) = 25mA Yellow: V_F (typical) = 2.1V, I_F (max) = 30mA Green: V_F (typical) = 2.0V, I_F (max) = 25mA	RS Components: Blue: 466-3548 Red: 228-5972 Orange: 228-5994 Yellow: 228-6010 Green: 228-6004
D1–D24 (Hardware)	24	LED mounting bezel, 5mm	RS Components 262-2999
SW1	1	Toggle switch, panel mount, 2A rated	RS Components 710-9674
Hardware	1	Stripboard, 0.1" (2.54mm) hole pitch, 37 holes wide by 24 tracks high (cut to a smaller size; see text)	—
Hardware	1	8-pin DIL socket	—
Hardware	2	16-pin DIL socket	—
Hardware	1	8-pin SIL pin header	—
Hardware	1	PP3 battery clip and lead	RS Components 489-021 (pack of 5)
Hardware	1	AA battery holder (three AA batteries)	Maplin YR61R
Hardware	3	AA battery (1.5V)	—

Chapter 9 ■ Driving Multiple LEDs from a Single IC Output

PARTS LIST			
Code	Quantity	Description	Supplier and Part Number
Hardware	1	Enclosure: translucent blue 3.94" (100mm) × 1.97" (50mm) × 0.98" (25mm)	RS Components 415-2745 (Mfr: Hammond 1591ATBU)
Hardware	—	LED display enclosure (see text), 2mm clear acrylic sheet, multicolor interconnecting wires, M3 standoffs, M3 nuts, self-adhesive pads, cable ties, cable tie bases	—

*Note: If you use LEDs that have different V_F and I_F values to those specified in the parts list, then you may need to alter the resistance and wattage values of these LED series resistors. Please refer to Chapter 2, which explains how to do this.

Deciding Which LED Enclosure to Use

You will soon see that I decided to use a slightly unconventional enclosure to display the 24 LEDs. After munching through a box of lovely chocolates while building some of the electronic projects in this book (it's hungry work, this electronics business!), I was about to throw away the plastic box that they came in when I realized that this would make an ideal housing for this project. The "chocolate box" is approximately 9 × 9 inches, is made from clear plastic, and has a base and a lid. It also has a plastic tray insert that was used to hold the 24 chocolates in place (before I ate them all), and this insert has a shiny metallic appearance, which is also useful for creating 24 individual reflectors to boost the light output of each LED. You can either make your own LED display enclosure or buy a premade enclosure with a clear lid (sold in a variety of sizes and styles by many component suppliers). I am going to describe how I built my project using the chocolate box, but this should give you some ideas about the type of enclosure you may want to use to build your project.

Stripboard Layout

The first thing you need to do is build the stripboard that contains all the driver circuitry, the layout for which is shown in Figure 9-3. As specified in the

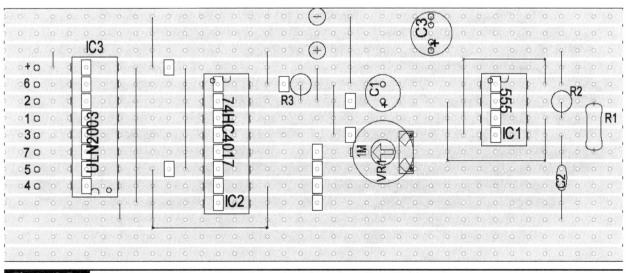

Figure 9-3 The stripboard layout for the color-changing disco light project

parts list table, the stripboard size is 36 holes wide by 15 tracks high; I used a larger piece of stripboard and cut it down to size. The eight pins positioned next to IC3 are part of an eight-way SIL pin header, which is included to make it easier to solder the remote LEDs to the board. Also note that you need to make 29 track cuts before you start to solder the components in place; these are shown as white rectangular blocks in Figure 9-3.

NOTE Please refer to Chapter 1 for soldering tips and techniques and for generic stripboard building guidelines.

How to Build and Test the Board

Because my chocolate box was not deep enough to accommodate all 24 LEDs, the stripboard, and the batteries, I decided to use a separate enclosure to house the circuit board, which I then mounted onto the back of the chocolate box. The Hammond enclosure that I used to house the stripboard has screw fixings in each corner, so I had to cut the corners off the stripboard to enable it to fit in place. If you decide to take a different approach and forego the separate enclosure, you may be able to fit the stripboard inside your LED display enclosure, in which case you may not need to cut the corners off the stripboard.

Build the stripboard carefully by following the stripboard layout shown in Figure 9-3. You should end up with a finished board that looks like the one shown in Figures 9-4 and 9-5. Eventually there will be eight cables coming from the LED display; you will solder these to the 8-pin header, which is shown on the left side of the layout diagram.

The next thing to do is to test the board to make sure that it works in the correct manner. Once you have made the usual visual checks of the board, apply a 4.5V supply and check that the power rail pins of each of the three IC sockets reads as expected using a multimeter. Now remove the power and insert IC3 only, making sure that it is fitted in the correct orientation; you will see that pin 1 is facing toward the bottom of the board, unlike IC1 and IC2. Reapply the power and fit a jumper wire between pin 16 and pin 1 of the DIL

Figure 9-4 How the finished stripboard should look

Chapter 9 ■ Driving Multiple LEDs from a Single IC Output

Figure 9-5 A close-up of the circuitry around the 555 timer

socket of IC2, as shown in Figure 9-6. This applies a positive supply to pin 7 of IC3 and should produce a negative supply output at pin 10 of IC3.

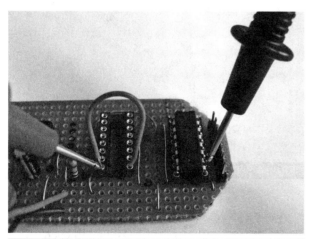

Figure 9-6 Testing that the outputs of IC3 work as expected

If this test works, then you can move the jumper link along to activate each of the base connections of IC3 in turn, to make sure that their corresponding collector outputs go low. Always keep one end of the jumper cable connected to pin 16 (+) of IC2 and move the other end along pins 1, 2, 3 (see Figure 9-7), 4, 5, 7, and 10 of IC2's DIL socket, checking that IC3 pins 10–16 go low accordingly.

If you are happy with the operation of IC3, remove the power from the board. Before you install IC1 and IC2 it is recommended that you remove the power and make sure that C3 is discharged; you can do this by connecting a 5mm red LED complete with a 470Ω series resistor across the supply rails of the stripboard. This can be done by connecting a leg of the series resistor to pin 8 of IC1's DIL socket (+ rail); the other leg of the series resistor is then connected to the anode of

104 Brilliant LED Projects

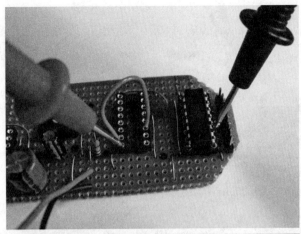

Figure 9-7 Check each IC3 output in turn by moving the jumper cable along each pin.

the LED, and finally the cathode leg of the LED is connected to pin 1 of IC1's DIL socket (− rail). Any charge left in C3 will illuminate the LED briefly; when the LED is no longer illuminated, C3 is discharged and it is safe to fit IC1 and IC2. You can now turn the variable resistor VR1 fully counterclockwise to its slowest position, insert IC1 and IC2 into their DIL sockets, and reapply the 4.5V supply. Use your multimeter to check that each of the seven header pins (numbered 1–7) next to IC3 go low briefly on a regular basis as the decade counter activates each one in turn. Once you are satisfied that the circuit operates correctly, it is time to build the LED display.

How to Build the LED Display

This is where you can be creative and decide how you want your LED flashing sequence to look. You don't need to copy my design if you don't want to, but there are some facts to keep in mind and some rules to follow when designing your own LED display:

- There are seven stages to the sequence.
- Each of the seven stage outputs of the ULN2003 can be connected to a maximum of four LED cathodes.
- Each of the 24 LED anodes needs to have its own series resistor soldered to it.
- The other ends of the 24 LED resistors need to be connected together, creating the common + connection that needs to be fed back to the stripboard (and soldered to the + pin of the pin header).
- You should end up with eight cables from the LED display feeding back to the stripboard: one common (+) connection and seven individual (−) connections.

I planned my LED display by writing the sequence numbers 1 through 7 on pieces of masking tape and then moving them around in different configurations on the chocolate box tray so that I could visualize what the sequence would look like. The numbers on the actual tray (not those on the tape) in Figure 9-8 show how my layout ended up after I played around with the pieces of masking tape.

Figure 9-8 How I designed the LED sequence before wiring up the LEDs

As you can see, stage 1 of the sequence illuminates a bank of four LEDs, one in each corner. I chose four blue LEDs for this stage.

Stage 2 of the sequence illuminates a bank of four red LEDs that forms another square within the stage 1 square. Stage 3 of the sequence illuminates four yellow LEDs in the center of the tray. You should be able to see how the sequence continues for stages 4 through 7 by following the numbers in Figure 9-8. The final visual effect that this produces is a sequence of squares that moves in and out, creating an ideal disco lighting effect. Table 9-1 shows how I connected the LEDs together and lists the resistor values that I used for each of the different color LEDs. The series resistors for the blue and red LEDs have a higher value than the others because this helps to match the brightness of the other LEDs. If you use different types of LEDs from those that I used, then you may need to alter the series resistor values accordingly.

The following instructions are based on how I built my display using the chocolate box enclosure. You'll need to adapt these instructions to suit the needs of your own enclosure.

TABLE 9-1	LED Configuration Used for the Project Display	
Stage Number	LED Color and Quantity	Series Resistor Value and Quantity
1	Blue (4)	180Ω (4)
2	Red (4)	180Ω (4)
3	Yellow (4)	150Ω (4)
4	Green (4)	150Ω (4)
5	Orange (4)	150Ω (4)
6	Red (1) Orange (1)	180Ω (1) 150Ω (1)
7	Green (1) Yellow (1)	150Ω (1) 150Ω (1)

CAUTION Take extra care when drilling into acrylic and thin plastic, as it can shatter and split easily. Use a battery-powered drill with a sharp drill bit on a very slow speed to drill pilot holes, and then use the larger drill bits on a slow speed to produce the desired hole size. Always wear safety goggles and take your time. Do not be surprised if you end up with a few cracks and splits in the plastic, which may be unavoidable (you will notice that my display has a few).

If you follow my idea and use something similar to the inner tray as reflective material, make sure that the material you use is an insulator and does not conduct electricity. This could be something as simple as an empty egg carton sprayed with a non-metallic silver paint.

1. Cut a piece of 2mm clear acrylic sheet to a size that is small enough to fit inside your display enclosure. Mark the position of the 24 LEDs on the acrylic sheet. I used the layout of the chocolate tray insert to mark the required positions of the LEDs on the acrylic. Carefully drill the holes for the LEDs (making sure that they are big enough to accommodate the required size for the LED holders) and also drill four smaller holes to allow the acrylic to be secured to the main enclosure. Then insert four M3 screws and secure them into the acrylic using M3 stand-offs (see Figure 9-9). You then need to carefully drill four holes into the base of the LED enclosure to suit the four acrylic mounting points, plus a fifth hole for the cable entry.

2. Fit the 24 LEDs into the holes using the LED holders, and then bend the leads of the LEDs over, being careful not to put stress on the end that goes into the LED housing. You can hold the base of the LED leads firmly using pliers and then bend the leads over by hand. Figure 9-10 shows the LEDs fitted and their leads bent over at 90 degrees.

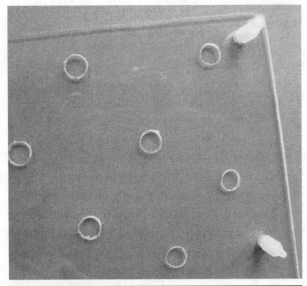

Figure 9-9 Drill the acrylic with the LED positions and insert the stand-offs.

Figure 9-11 Sand down the inside of the lid.

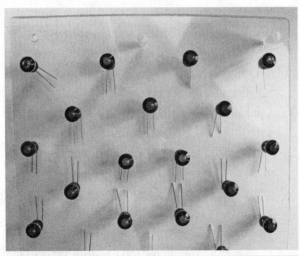

Figure 9-10 Insert the LEDs into the acrylic and bend their leads over.

3. Sand down the inside of the LED enclosure lid using sandpaper. This creates a frosted effect (see Figure 9-11) and helps to make the light from the LEDs look bigger and brighter in the dark.

4. Solder all the LEDs together in the required manner. This process may take a few hours if you do it carefully, but the final result makes it worthwhile. Here are the steps:

a. Connect multiple LEDs into a bank by soldering their cathode (−) connections together. I started by connecting the four middle, yellow LEDs together (stage 3 in Figure 9-8). Solder this common cathode connection for the four LEDs to a piece of wire that is long enough to reach from the display to the stripboard. I chose to use a different color of wire for each LED bank, corresponding to the color of the LEDs in that bank, to help me identify which color LED bank the wire is connected to. Thus, I used yellow wire for the middle bank. You could use individual strands of ribbon cable for the flying leads so long as the cable that you use is able to support the total current drawn by each bank of LEDs.

b. Solder a resistor to the anode (+) lead of each of the four LEDs. Join the other end of the four resistors together, as shown for the middle bank in Figure 9-12, to form the common anode connection that you'll connect to the stripboard. In Figure 9-12, the common anode resistor connection for the middle bank of LEDs still needs to have a flying lead connected to it. As you can see, Figure 9-12 also

Chapter 9 ■ Driving Multiple LEDs from a Single IC Output 107

Figure 9-12 Use colored cables to help you identify each color bank.

shows part of the wiring for the stage 1 bank of (blue) LEDs. Because these LEDs are not close enough together to join the ends of the resistors directly, each resistor requires a flying lead connected to it. I used a white cable for this purpose. This common + connection is soldered to all 24 resistors.

c. Test the bank of LEDs by connecting the negative connection of a 4.5V battery to the common cathode (−) connection of the bank of four LEDs and connecting the positive connection of the 4.5V battery to the common anode resistor connection. All four LEDs should illuminate. (Testing each bank of LEDs in this manner helps to reduce the amount of fault-finding needed when you finally switch on the finished project.)

d. Repeat the preceding three steps for each bank of four LEDs (or two, in the case of stages 6 and 7).

e. Affix the eight flying cables (one for each of the seven banks and one for the common anode resistor connection) to the acrylic using a self-adhesive pad and a

cable tie. The final layout of my LED display is shown in Figure 9-13 (bottom, solder side) and Figure 9-14 (display side).

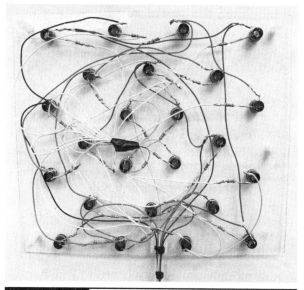

Figure 9-13 The solder side of the display

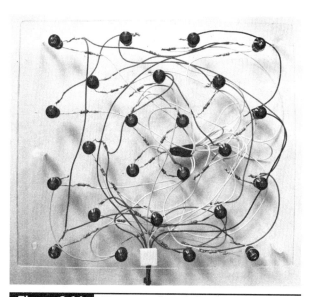

Figure 9-14 The completed display side

5. Mount the completed acrylic LED display onto the rear of the main display enclosure using the M3 screws and nuts, as shown in Figure 9-15. Then use double-sided adhesive to affix the AA battery holder and the circuit board enclosure onto the back of the display enclosure, as shown in Figure 9-16.

Figure 9-15 Mount the acrylic onto the main display enclosure using the M3 screws.

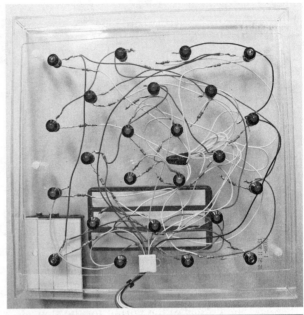

Figure 9-16 Affix the battery holder and stripboard enclosure to the back of the display.

6. Feed the eight flying LED cables through the hole in the stripboard enclosure and solder the connections in the correct order to the pin headers that are positioned next to IC3. Make sure that you connect the common anode connection from all 24 LEDs to the top header pin marked + on the stripboard layout diagram shown in Figure 9-3. Solder the battery cables to the board, making sure that the positive lead is wired via the power switch SW1. Figures 9-17 and 9-18 show the results from my project.

Figure 9-17 Solder the LED cables to the pin headers.

Figure 9-18 The soldering is now complete.

7. At this stage, before you fit the lid, you might want to connect the battery and switch the unit on to make sure that the LED sequence works. You should see an eye-catching sequence of LEDs. If you have followed my design, then you should see a mini-animation of a square box of four LEDs moving in and out of the center of the display. If you have been careful and have been fault-finding and testing as you have been building the project, it should work the first time you try it.

8. Fit the lid of the display enclosure. (Before I did this, I made holes in each segment of the metallic-looking chocolate box tray to allow the light from each of the LEDs to shine through, and then I placed this over the LED display, as shown in Figure 9-19.) Apply a length of clear adhesive tape along each edge to hold the lid in place.

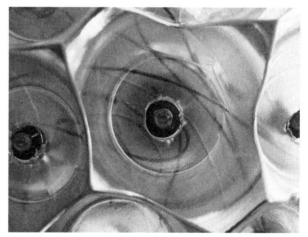

Figure 9-19 Make holes in the tray to allow the LEDs to shine through.

Figures 9-20 and 9-21 show my completed project. You will also notice that I drilled a small hole in the lid of the stripboard enclosure, which allows me to adjust the variable resistor and alter the speed of the display.

Figure 9-20 The back of the finished project, showing the reflector tray fitted

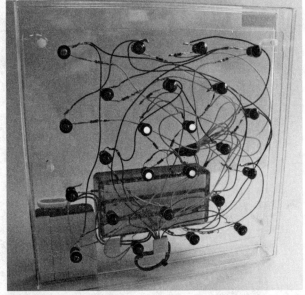

Figure 9-21 The front of the LED display without the reflector tray and lid fitted

Time to Disco

The disco lights are best viewed in the dark (while listening to the *Saturday Night Fever* soundtrack, naturally). The visual effect produced by the sanded lid and the individual reflectors is quite stunning. Try adjusting the variable resistor to alter the speed of the display. You will also notice that, at full speed, the LEDs are flashing so fast that it almost looks as though all 24 LEDs are illuminated at the same time. The projects in Part Three investigate this effect and put it to good use.

CHAPTER 10

LED Binary Ripple Counter

The project in this chapter demonstrates the operation of a binary ripple counter IC by building an experimental board that shows how each output of the IC operates, using a single LED indicator per output. You could use the board for educational purposes to show how to count in binary, or you could build multiple boards to produce a mock-up of a mainframe computer that looks as though it is performing some really complex calculations!

If you previously built the color-changing light box in Chapter 6, you have already met the 4060 14-stage binary ripple counter IC. That project used just three of its outputs to change the color of an RGB LED. This chapter first delves into the operation of this IC and its 74HC version in a little more detail, and then shows you how to build the experimental board shown in Figure 10-1.

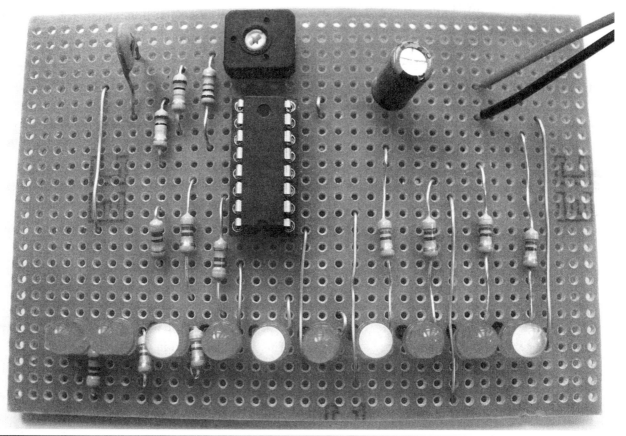

Figure 10-1 LED binary ripple counter project

The 4060 and 74HC4060 Binary Ripple Counters

The 4060 CMOS device is a 14-stage binary ripple counter that also has the benefit of including its own internal oscillator, which means that there is no need to provide additional clock circuitry. This is a nice feature because it helps you to keep the overall component count down when designing projects. The IC's internal oscillator can be generated either by a resistor/capacitor (RC) network only or, alternatively, by both an RC network and a crystal if accurate timing is required (for example, to create the timing circuitry for a digital clock). In this project, the accuracy of the oscillator isn't important, so you are going to use an RC network sans a crystal.

The color-changing light box project in Chapter 6 used a standard 4060 CMOS device. That project included separate transistor driver circuitry because the current-sourcing capabilities of the 4060 device are limited. In this project, you will use the 74HC variation of the 4060 device, which allows each output to drive up to 25mA (see the caution below); this means that you do not need to use additional transistor circuitry to drive the required single LED per output.

CAUTION Do not use a standard 4060 device in this project, as the current drawn from the LEDs could damage the IC. It should also be noted that the LED series resistors used in this project have been set to limit the current to each LED to less than 5mA. This means that the overall current consumption of the IC does not exceed the maximum recommended DC supply current, which is stated as 50mA on the manufacturer's datasheet.

Project 9
LED Binary Ripple Counter

Have a look at the project specifications for the LED binary ripple counter. I originally designed this project to show how to count in binary but it can be modified for other purposes.

> **PROJECT SPECIFICATIONS**
> - Ten individual 5mm red LEDs are used to display a binary counting sequence.
> - The result of the project can be used as an educational tool or to make a sci-fi prop.
> - Operation requires a single chip, with no additional clock IC required.
> - The supply voltage is 4.5 volts.

How the Circuit Works

Figure 10-2 shows the circuit diagram for the LED binary ripple counter project.

The circuit is fairly straightforward. You are powering the board using a 4.5V supply (in the form of three 1.5V AA cells), and this powers the 74HC4060 IC. Do not forget that the maximum supply voltage allowed for this IC is 6V, unlike the standard 4060 device, which has a higher upper voltage limit. The speed of the IC's internal clock is determined by the values of R1, R2, C1, and VR1. The datasheet for the 74HC4060 device that I used shows a formula and offers some component value guidelines, which helps you to calculate the frequency of the oscillator circuit. I used this information to help me to identify, and experiment with, a few different component values. This finally resulted in me choosing the components used in this project.

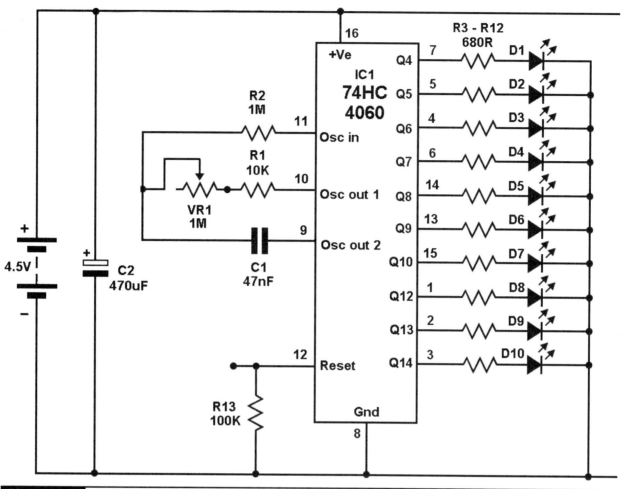

Figure 10-2 The LED binary ripple counter circuit diagram

NOTE If you are looking for more accurate timings when using this device, use a crystal and an RC network to drive the oscillator rather than just an RC network. You can get more information about how a crystal can be connected to the IC by referring to the relevant manufacturer's datasheet.

In this project's circuit, the RC network produces a binary count frequency of approximately 1Hz when VR1 is set to its maximum setting of 1MΩ, and this speed increases when VR1 is adjusted to a lower resistance setting. When the internal oscillator operates in the circuit, it produces a binary count on its output pins (Q4 to Q14); the least significant bit of the binary count is output on pin 7 and the most significant bit is output on pin 3. The reset connection for IC1 (pin 12) is held low via R13 to allow the IC to count, and C2 is a decoupling capacitor that is positioned across the supply rail.

Parts List

NOTE The Supplier and Part Number column of the following table lists specific parts that I used in this project. Refer to the appendix for additional details about acquiring your parts.

The parts you'll need for the LED binary ripple counter project are listed in the following table.

PARTS LIST			
Code	Quantity	Description	Supplier and Part Number
IC1	1	74HC4060A 14-stage binary ripple counter with oscillator	RS Components 625-4871 (or similar)
R1	1	10KΩ 0.5W ±5% tolerance carbon film resistor	—
R2	1	1MΩ 0.5W ±5% tolerance carbon film resistor	—
R3–R12*	10	680Ω 0.5W ±5% tolerance carbon film resistor	—
R13	1	100KΩ 0.5W ±5% tolerance carbon film resistor	—
VR1	1	1MΩ miniature enclosed horizontal preset potentiometer (min. 0.15W rated)	—
C1	1	47nF ceramic disk capacitor (min. 10V rated)	—
C2	1	470μF 10V radial electrolytic capacitor	—
D1–10	10	5mm red LED V_F (typical) = 2.0V, I_F (max) = 30mA	RS Components 228-5972 (pack of 5)
Hardware	1	Stripboard, 0.1" (2.54mm) hole pitch, 37 holes wide by 24 tracks high	—
Hardware	1	16-pin DIL socket	—
Hardware	1	PP3 battery clip and lead	RS Components 489-021 (pack of 5)
Hardware	1	AA battery holder (three AA batteries)	Maplin YR61R
Hardware	3	AA battery (1.5V)	—

* Note: If you use LEDs that have different V_F and I_F values to those that are in the parts list, then you may need to alter the resistance and wattage values of these LED series resistors. Please refer to Chapter 2, which explains how to do this, and also remember that you should be using a value of less than 5mA for the value of I_F in your calculations for this project.

Stripboard Layout

Figure 10-3 shows the stripboard layout for the LED binary ripple counter.

You need to make 17 track cuts in total (these are shown as white rectangular blocks), eight underneath IC1's DIL socket and nine next to the ten LEDs.

How to Build and Test the Board

NOTE Please refer to Chapter 1 for soldering tips and techniques and for generic stripboard building guidelines.

Build the stripboard following the diagram shown in Figure 10-3. You should end up with a stripboard layout that looks like the one shown in Figure 10-4. I recommend that you solder in place last the ten LEDs, which allows you clear access to solder the interconnecting links and resistors into the holes near to where the LEDs are going to be positioned before they are soldered in place.

Figure 10-5 shows a close-up of the area around the ten LEDs, when looking at the board upside down so that the LEDs are at the bottom of the board. This is the position in which you need to view the board to see the binary count sequence properly.

Chapter 10 ■ LED Binary Ripple Counter 115

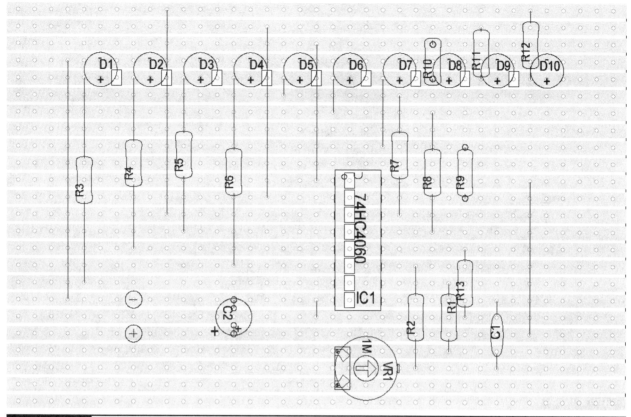

Figure 10-3 Stripboard layout for the LED binary ripple counter

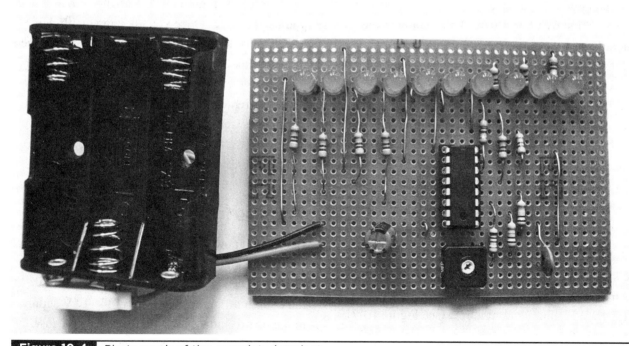

Figure 10-4 Photograph of the completed project

Figure 10-5 The layout around the ten LEDs

Before you fit IC1 into its DIL socket, it's worth performing some tests that will help you to avoid damaging the IC in the event of any constructional errors. First of all, apply the 4.5V supply to the board. At this stage, none of the LEDs should be illuminated; if any are lit, then remove the battery and check the board over to spot any mistakes. If all the LEDs are off, then make sure that pin 16 of the DIL socket sits at +4.5 volts compared to pin 8 of the DIL socket using a multimeter. If this is the case, then fit a small jumper wire between pin 16 (+) and pin 7 of the DIL socket, as shown in Figure 10-6, and the rightmost LED (D1) should illuminate.

Next, keep one end of the jumper wire connected to pin 16 and move the other end of the wire along each of the output pins 7, 5, 4, 6, 14, 13, 15, 1, 2, and 3 one at a time (see Figure 10-7).

This should prove the operation of each of the LEDs D1 to D10 in turn from right to left when the LEDs are at the bottom of the board. If at any point the relevant LED does not illuminate, or if multiple LEDs illuminate, then there is a fault that you need to find. Once you are happy with the operation of the LEDs, remove the supply but keep the wire link between pin 16 and pin 7—this helps to discharge capacitor C2 through D1 when the battery is removed. When D1 no longer illuminates, remove the wire link and then carefully insert IC1 into its DIL socket, making sure that it is correctly oriented.

Counting in Binary

Before you power up the board, turn the variable resistor VR1 so that it is fully clockwise, and then turn the board upside down so that the LEDs are at the bottom of the board. Connect the 4.5V supply to the board, and this should immediately cause the LEDs to show a binary count sequence. This should start off in a similar manner to the pattern

Chapter 10 ■ LED Binary Ripple Counter 117

Figure 10-6 Connect pin 16 to pin 7 to illuminate D1.

Figure 10-7 Continue to test the output pins.

depicted in Table 10-1, which shows how the first four LEDs (D1 to D4) operate; a value of 1 denotes that the LED is illuminated, and a value of 0 means that it is off. If you find that the circuit does not operate as expected, then either there is a problem in the RC network or IC1 may be faulty. In this instance, you will need to follow the circuit diagram and fault-find the stripboard to find out where the problem lies.

Now slowly adjust VR1 counterclockwise, and the speed of the binary count starts to increase as the speed of the internal oscillator increases.

Turning VR1 fully counterclockwise produces a very fast binary clock count. Also note that if you briefly connect the positive supply connection to reset pin 12, it will reset the binary count to 0.

You will notice that Q11 is missing from this device, which means that a true, uninterrupted binary count to 128 is visible on the LEDs connected to pins Q4 to Q10 (D1 to D7). Once the count exceeds 128, part of the binary sequence goes missing for a while because you cannot see the value of Q11. As mentioned at the beginning of the chapter, you could use this experimental board for educational purposes as a visual aid when demonstrating how to count in binary, or you could use it as a prop when making a mock "computer system."

Try winding the variable resistor to its halfway point, so that the clock speed is fairly slow, and covering up all the LEDs apart from D1; watch how fast it flashes. Now cover up all the LEDs apart from D2. What do you notice? Try this again by covering up all LEDs apart from D3. What do you see? What you should notice is that D1 is flashing at a regular rate of approximately two times per second, and when you then look at D2 in isolation, you should notice that the flash rate of this LED is slower than that of D1. And, again, when you look at D3 in isolation, you will see that the speed of D3 is slower than that of D2.

The project in the next chapter utilizes the digital outputs of this IC in a slightly unconventional way, producing a totally different lighting effect that has nothing at all to do with binary counting or computers.

TABLE 10-1	Typical Binary Count for LEDs 1 to 4			
Count	D4 (Q7)	D3 (Q6)	D2 (Q5)	D1 (Q4)
0	0	0	0	0
1	0	0	0	1
2	0	0	1	0
3	0	0	1	1
4	0	1	0	0
5	0	1	0	1
6	0	1	1	0
7	0	1	1	1
8	1	0	0	0
9	1	0	0	1
10	1	0	1	0
11	1	0	1	1
12	1	1	0	0
13	1	1	0	1
14	1	1	1	0
15	1	1	1	1

CHAPTER 11

Flickering LED Candle

THE PROJECT IN THIS CHAPTER SHOWS YOU HOW TO build a flickering LED candle. The end result is a small "candle" with a fairly realistic-looking light output, as shown in Figure 11-1.

Figure 11-1 The flickering LED candle

This sequencer circuit project drives only a single LED output. Although this does not follow the convention for the sequencer projects in this part of the book, this circuit demonstrates how you can use the binary ripple counter IC from Chapter 10 and modify its output sequence to produce something totally different from what the IC was originally intended for.

Project 10
Flickering LED Candle

The LED candle that you create in this project could be used, for example, inside a carved pumpkin on Halloween, as a nightlight, or for theatrical lighting effects.

PROJECT SPECIFICATIONS

- The result of this project is a flameless electronic candle.
- The candle has a realistic-looking flickering "flame" effect.
- The candle uses a warm white LED for light output to approximate the color of a candle flame.
- The circuit design is compact and inexpensive, using a single IC and nine other components.
- The supply voltage is 4.5 volts.

How the Circuit Works

The circuit diagram for the flickering LED candle is shown in Figure 11-2.

119

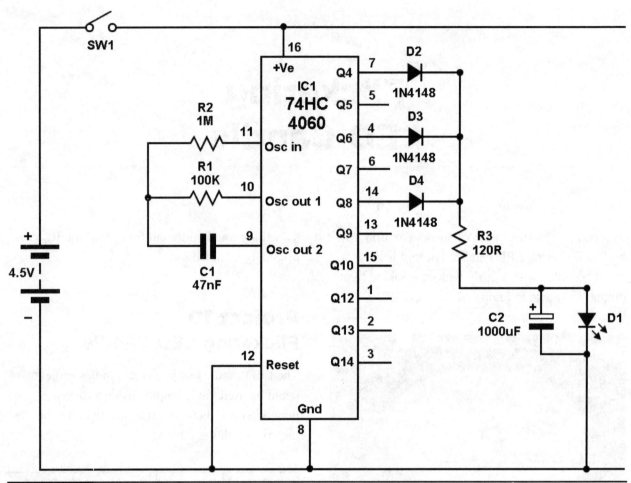

Figure 11-2 Flickering LED candle circuit diagram

The circuit is powered by a 4.5V supply in the form of three AAA batteries. This drives the 74HC4060A 14-stage binary counter (IC1), which was introduced in Chapter 10. As explained in Chapter 10, the ten outputs of this device produce a binary count, the speed of which is controlled by the IC's internal clock circuit. In this circuit, the speed of IC1 utilizes an RC timing configuration using resistors R1 and R2 and capacitor C1. The clever bit about the present project is that if you increase the speed of the IC's internal oscillator and combine three of the binary outputs, you can produce a single-pulsed digital output, which will create a flickering effect when fed into an LED. The three digital outputs are combined into a single output via the three signal diodes (D2 to D4).

Including these diodes is important because they prevent each of the digital outputs from feeding back into the others, which could damage the IC if the diodes were not included in the circuit.

When designing the circuit, I discovered by trial and error that the most realistic digital output combination to produce a candle flicker is Q4, Q6, and Q8. Because each of the binary outputs switches on and off at different speeds, the combination of these three signals produces a modified digital output. When this digital output is fed into a single LED (D1), it produces the flickering effect required. The three signal diodes (D2–D4) feed this signal into this LED via series resistor R3, which allows less than 15mA to flow through the LED used. Initially I considered using a yellow LED for this project, but I ended up using a warm white LED instead, which produces a realistic-looking candle light.

You will also notice that there is a 1000μF electrolytic capacitor (C2) wired in parallel across the LED; the purpose of this capacitor is to smooth the combined digital output signal, which produces a much more realistic flicker. The other byproduct of this electrolytic capacitor is that when you switch the candle off, the small charge that remains in the capacitor discharges through the LED and keeps it lit very dimly for a few seconds. This makes it look like the candle is glowing once it has been extinguished, just like the real thing.

Parts List

NOTE The Supplier and Part Number column of the following table lists specific parts that I used in this project. Refer to the appendix for additional details about acquiring your parts.

Here are the parts you'll need for the flickering LED candle project.

		PARTS LIST	
Code	Quantity	Description	Supplier and Part Number
IC1	1	74HC4060A 14-stage binary ripple counter with oscillator	RS Components 625-4871 (or similar)
R1	1	100KΩ 0.5W ±5% tolerance carbon film resistor	—
R2	1	1MΩ 0.5W ±5% tolerance carbon film resistor	—
R3*	1	120Ω 0.5W ±5% tolerance carbon film resistor	—
C1	1	47nF ceramic disk capacitor (min. 10V rated)	—
C2	1	1000μF 10V radial electrolytic capacitor (see text)	—
D1	1	5mm warm white LED (LM520A) V_F (typical) = 3.3V, I_F (typical) = 20mA	RS Components 667-5846 (pack of 5)
D2–D4	3	1N4148 signal diode	—
SW1	1	Single-pole panel mount toggle switch, 2A rated	RS Components 710-9674
Hardware	1	Stripboard, 0.1″ (2.54mm) hole pitch, 25 holes wide by 9 tracks high	—
Hardware	1	16-pin DIL socket	—
Hardware	1	AAA battery holder (four AAA batteries; see text)	RS Components 512-3568
Hardware	3	AAA battery (1.5V)	—
Hardware	—	Enclosure, 2mm clear acrylic sheet, cable tie, double-sided adhesive tape (see text)	—

*Note: If you use an LED that has different V_F and I_F values to those that are in the parts list, then you may need to alter the resistance and wattage values of this LED series resistor. Please refer to Chapter 2, which explains how to do this.

Stripboard Layout

The stripboard layout for the flickering LED candle is shown in Figure 11-3.

The stripboard layout is fairly simple and shouldn't take too long to build. Note that you need to make 12 track cuts before you start to solder the components in place; these include cuts that sit underneath R1, R2, and C1. Track cuts are shown as white rectangular blocks in Figure 11-3.

How to Build and Test the Board

NOTE Please refer to Chapter 1 for soldering tips and techniques and for generic stripboard building guidelines.

Build the stripboard by carefully following the diagram shown in Figure 11-3. You should end up with a finished stripboard that looks like the one shown in Figure 11-4. Notice that capacitor C2 and LED D1 are positioned so that they both lay flat and that their hole positions differ slightly in

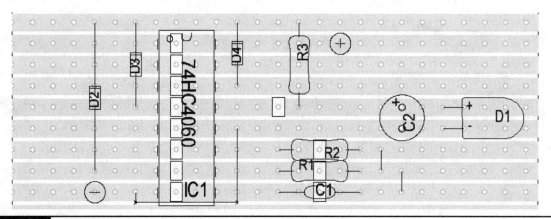

Figure 11-3 The stripboard layout for the flickering LED candle

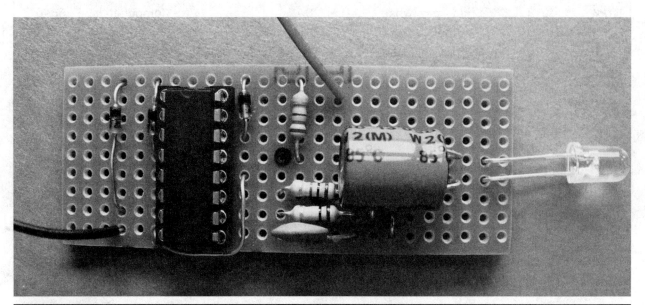

Figure 11-4 The completed stripboard

Figures 11-3 and 11-4. I did this to save space and to make sure that the LED was able to protrude through the top of the candle enclosure that I used.

The circuit needs to be powered using three AAA cells, rather than four, to save on space. I struggled to find a compact AAA battery holder that would hold just three batteries, so I ended up modifying a four-battery AAA battery holder by shorting out one of the battery compartments. I did this by carefully soldering a piece of battery wire across one of the battery compartments, as shown in Figure 11-5.

CAUTION Be careful when soldering each end of the wire to the battery connectors, because the heat of your soldering iron can quickly start to soften the plastic casing of the battery holder if it is left on the solder joint too long. Also make sure that the current rating of the wire is higher than the total current consumption of the circuit.

After you have soldered the wire into place, insert three AAA batteries into the three remaining spaces of the battery holder and check with a multimeter that the output voltage from the holder is 4.5 volts.

The component count for this circuit is fairly low, so you may be confident that it will work the first time; however, to be safe, you should make some initial tests before inserting IC1 into its

Figure 11-5 Solder a piece of wire into one of the four battery compartments.

socket, similar to the tests performed in Chapter 10. Initially, check that DIL socket pin 16 is positive compared to pin 8. If this is the case, you can then use a wire link to link pin 16 and each of the three output socket pins 4, 7, and 14 in turn, making sure that the LED illuminates every time one of these three output pins has a + voltage applied to it.

Once you are happy with your stripboard layout, insert IC1 into the DIL socket and then apply the 4.5V supply to the board. This should cause the LED to illuminate and flicker like a candle. If this does not happen, then remove the battery and check the board over again before you proceed. Figure 11-6 shows another close-up photograph of the completed stripboard from a different angle.

Figure 11-6 Close-up of the finished stripboard

The LED Candle Enclosure

You may already have some ideas about the type of enclosure that you want to use for your project. I decided to use an old plastic pill container, which I cleaned up and then drilled two holes into, one hole in the bottom of the container for the LED to shine through and another in the side of the container to accommodate the switch (see Figure 11-7).

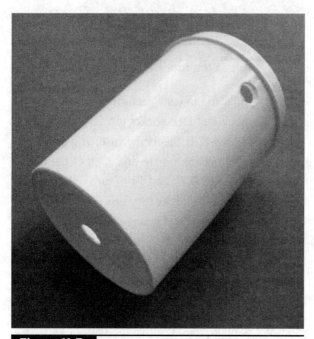

Figure 11-7 Drill two holes into the container.

Figure 11-8 The stripboard is secured to the acrylic using a cable tie.

Because of the limited space that I had to play with inside the pill container, I decided to create two separate compartments, one for the battery holder and one for the stripboard. There are two reasons for doing this: to stop the copper tracks of the stripboard from shorting out against the battery holder, and to help to keep the battery holder and the stripboard secure. I used a piece of shaped 2mm clear acrylic to create the compartment, as shown in Figure 11-8, and then I secured it to the lid of the container using double-sided adhesive tape. I attached the stripboard to the acrylic using a cable tie. I then cut the positive battery cable from the battery holder and soldered each end to the power switch, which I fitted through the hole in the side of the container. Notice in the finished result, shown in Figure 11-8, that the AAA battery holder sits in its own space underneath the stripboard on the other side of the acrylic.

All that you need to do now is to hold the lid of the container and carefully maneuver the battery holder and the stripboard into the container, being careful to ensure that the switch contacts do not come into contact with anything once everything is in place. You may also need to adjust the position of the LED so that when you close the lid of the container, the LED protrudes through the hole that you drilled into the base of the container. The lid of the container becomes the base of your candle. The completed project should look something like the LED candle shown in Figure 11-9.

To make your candle container look even more realistic, you could drip some thick glue or paint down the side of the container and let it dry to give the effect of molten wax.

Experimenting with the Circuit

The circuit works quite well as it stands, but you might want to increase the value of C2 to 2200μF to further smooth the amount of light flicker from the LED. Or you could increase the value of R1 to

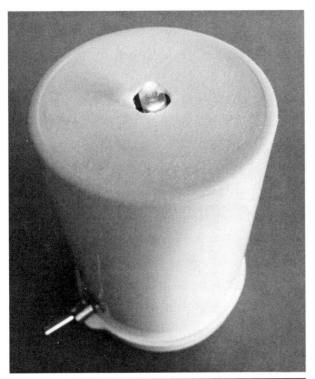

Figure 11-9 The finished LED candle

180KΩ, which would slow down the speed of the flickering effect. You could also build a number of these electronic candles with slightly varying oscillator speeds and mount them in various different candle-shaped containers. Switching them all on in the dark would produce a really spooky Halloween effect.

Alternative IC?

I originally designed the circuit around the 74HC4060 device because it has a greater output current capability than its standard 4060B counterpart. However, I have also experimented with this circuit using a standard 4060B CMOS device to see if it would work. It not only worked, but the flickering effect seemed to look slightly more realistic than that produced by the 74HC4060 device. I can only assume that there must be a difference in the way that these two devices switch from one binary state to another. In normal circumstances, the 4060B device should not be used to drive an LED, but in this circuit the LED drive current is sometimes being shared between two or three of the outputs, and they are switching at a fairly fast rate, so this may help to reduce the impact on the IC's outputs.

CAUTION The current output capabilities of the 4060B device are not really suited to driving an LED directly, and I have not tested the long-term effects of using this IC in this circuit, so if you decide to use the 4060B device instead of the 74HC4060 device, don't be surprised if you end up with a frazzled IC after a while!

CHAPTER 12

Introducing the PIC16F628-04/P Microcontroller: LED Scanner

THIS PROJECT EMULATES A LIGHTING EFFECT THAT you may have seen before. You may remember a popular TV show from the 1980s called *Knight Rider* that featured a talking car that had this type of lighting effect built into the front of its hood. A similar effect was used for the eye scanners of the Cylon Centurions in *Battlestar Galactica*. The LED scanner project comprises a line of eight LEDs, each of which illuminates sequentially in a way that makes it look as though a single LED scans from left to right and then right to left in a never-ending sequence. The other feature of this project is that as the LEDs scan from one side to the other, they leave a faint trail of slowly fading light, which disappears after a short period of time. This LED scanner effect could be used as an exciting addition to a children's toy, or it could be used to create an unusual LED badge. A photograph of the LED scanner project is shown in Figure 12-1. This is the first project in the book that incorporates a microcontroller at the heart of its circuit.

The PIC16F628-04/P Microcontroller

If you are not familiar with the PIC range of microcontrollers beyond the introduction provided in Chapter 1, there is some information coming up that explains how this type of integrated circuit operates and how they can be programmed. Before you jump ahead into programming, though, have a look at the key specifications of the PIC16F628-04/P microcontroller, which is the type of 8-bit device that I chose to use in this book. There are many different variants of PIC microcontrollers available with varying specifications and pin configurations; however, you will soon see that the PIC16F628-04/P is a really versatile device that can be used to create many different visually exciting LED projects.

Figure 12-1 The LED scanner

Key Specifications of the PIC16F628-04/P PIC Microcontroller
- 18-pin DIL package
- 3V to 5.5V supply range
- Input/output (I/O) pins split into Port A and Port B, which are configurable via software
- Up to 25mA sink/source current available from an output pin (which is enough to power a typical LED directly)—this is an absolute maximum rating and is subject to other parameters, see the note below*
- Internal 4-MHz clock (or can be controlled by an external crystal)
- 2048-word flash program memory (which is plenty of memory even for the more complex projects)
- More detailed specifications for this device can be found on the manufacturer's datasheet (available from www.microchip.com)

*Note: If you need to recalculate the value of any of the LED series resistors for the PIC Microcontroller projects in this book, make sure that you use an I_F value that is lower than 20mA. (More specific details are provided in each chapter.) If you are redesigning any of the projects, you should also refer to the manufacturer's datasheet to ensure that you are not exceeding the maximum total current and power dissipation capabilities of this device. |

PROJECT SPECIFICATIONS
- The display comprises a single row of eight large 8mm LEDs.
- The visual effect is a single LED that scans from side to side.
- The scanning LED leaves a slowly fading light trail behind it.
- The circuit is driven by the PIC Microcontroller.
- The speed of the LED scanning is controlled via two push-to-make switches.
- The supply voltage is 4.5 volts.

Project 11
The LED Scanner

The key project specifications that I used to design the LED scanner are outlined next.

How the Circuit Works

If you've completed some of the earlier projects in this book, you probably realize that I could have built this project using a 555 timer and one or two 4017 decade counters. However, I like to experiment with alternative circuit methods, and one of my key objectives for this project was to try and reduce the component count to a minimum. I also wanted to introduce you to the PIC Microcontroller. Thus, I decided that the best way to make this circuit is to use a PIC Microcontroller. This means that you have to use only a single IC and a handful of resistors and capacitors. The PIC Microcontroller used in this project is the PIC16F628-04/P device, which will also be used in some future projects.

The circuit diagram shown in Figure 12-2 demonstrates that the use of the microcontroller vastly simplifies the circuit layout. The 4.5V battery powers IC1 via pins 5 and 14, and resistor R1 feeds a positive signal into pin 4, which allows

The benefit of using this type of device in your projects is that you can write your own program (also known as *firmware*), which allows you to configure the inputs and outputs of the microcontroller so that they switch on and off in the manner that you require. This could be as simple as emulating the operation of the 74HC4017 decade counter IC, which was used in the sequencer project in Chapter 8; however, as you will see in later chapters, this device can produce even more interesting and complex LED effects.

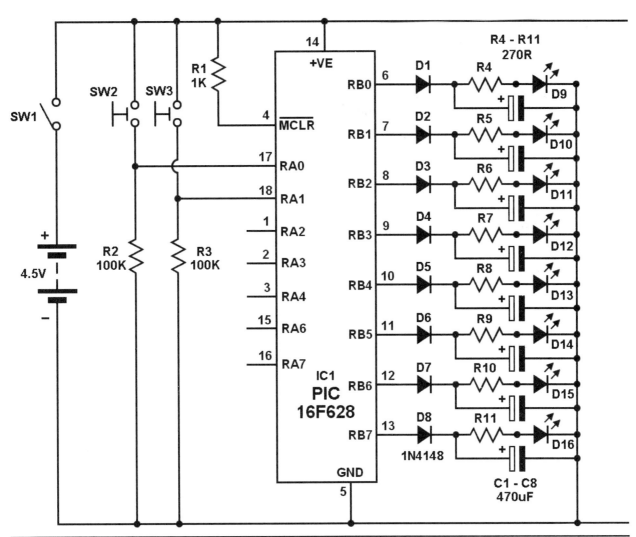

Figure 12-2 Circuit diagram for the LED scanner

the internal program to run when IC1 is powered up. This project does not require precision timing, so you will use the IC's internal 4-MHz clock, which means that no external clock circuitry is required, simplifying the circuit layout further. All eight I/O ports of Port B are configured as outputs in the program, and these ports are connected directly to the eight LEDs, D9 to D16, via eight signal diodes, D1 to D8. Resistors R4 to R11 are series current-limiting resistors for the eight LEDs.

You will notice in Figure 12-2 that each LED also has a 470μF electrolytic capacitor straddling it and its series resistor; it is this resistor/capacitor configuration that allows each LED to remain lit for a short period of time and creates the fading-light effect. If output pin 6 is made to go high, this positive voltage feeds through signal diode D1 and illuminates LED D9, at the same time charging up capacitor C1. If pin 6 is then made to go low, the LED would normally switch off immediately, but because C1 retains a charge, it discharges through series resistor R4 and keeps the LED illuminated for a short period of time. The purpose of the signal diode is to prevent the capacitor from discharging into the output pin of IC1, which could damage it. Now imagine each of Port B's output pins activating briefly, one at a time, starting at pin 6 and working down to pin 13, and then back up again to pin 6 in a never-ending loop. This scanning, in conjunction with each LED's resistor/capacitor network, creates the visual effect that you are looking for in this project.

You also want to control how fast the LEDs move from one side to the other. In previous projects, a variable resistor was used to control this. This project uses two momentary push-to-make switches, SW2 and SW3, instead. These switches are connected to Port A pins 17 and 18, which are configured as inputs in the program. Resistors R2 and R3 normally keep these pins low, and pressing SW2 or SW3 creates a positive signal on either pin, which is detected via the software. Pressing SW2 increases the speed, and pressing SW3 decreases the speed. The other I/O pins of Port A are configured as outputs in the program to avoid spurious triggering of IC1, and these pins are left unconnected.

Parts List

> **NOTE** The Supplier and Part Number column of the following table lists specific parts that I used in this project. Refer to the appendix for additional details about acquiring your parts.

The parts you'll need for the LED scanner project are listed in the table below.

PARTS LIST

Code	Quantity	Description	Supplier and Part Number
IC1	1	PIC16F628-04/P Microcontroller	RS Components 379-2869 (Mfr: MicrochipTechnology Inc. PIC16F628-04/P)
R1	1	1KΩ 0.5W ±5% tolerance carbon film resistor	—
R2/R3	2	100KΩ 0.5W ±5% tolerance carbon film resistor	—
R4–R11*	8	270Ω 0.5W ±5% tolerance carbon film resistor	—
C1–C8	8	470μF 10V radial electrolytic capacitor	—
D1–D8	8	1N4148 signal diode	—
D9–D16	8	8mm red LED V_F (typical) = 1.85V, I_F (typical) = 20mA	RS Components 577-718
SW1	1	Single-pole panel mount toggle switch, 2A rated	RS Components 710-9674
SW2, SW3	2	Single-pole normally open panel mount switch (100mA)	RS Components 133-6502
Hardware	1	Stripboard, 0.1" (2.54mm) hole pitch, 37 holes wide by 24 tracks high	—
Hardware	1	18-pin DIL socket	—
Hardware	1	AA battery holder (three AA batteries)	Maplin YR61R
Hardware	3	AA battery (1.5V)	—
Hardware	1	PP3 battery clip and lead	RS Components 489-021 (pack of 5)
Hardware	—	Flexible interconnecting wire	—

* Note: If you use LEDs that have different V_F and I_F values to those that are in the parts list, then you may need to alter the resistance and wattage values of these LED series resistors. Please refer to Chapter 2, which explains how to do this, and use a value of 10mA to 12mA for I_F in your calculations for this project (also note that changing these resistor values will alter the time that each LED stays illuminated during the scanning process)..

Stripboard Layout

The stripboard layout for the LED scanner is shown in Figure 12-3.

No enclosure is shown for this project because you might have specific ideas about where you want to house the LED scanner board. Similarly, the stripboard layout shows all the components mounted on the stripboard except the switches, because you might decide to mount the stripboard in a separate enclosure from that of the switches. Note that you need to make 23 track cuts (shown as white rectangular blocks) before you start to solder the components in place, including those underneath IC1 and around each of the capacitors and LEDs. Figure 12-3 also shows four 3mm holes in the stripboard, identified by the four white circles with a + sign in each corner; these holes are optional depending on how you intend to mount the board. You will also notice that there are no track cuts around the hole positions, which means that you should use insulated fixings, such as the nylon screws that have been used in previous projects (such as Chapter 7).

How to Build and Test the Board

NOTE Please refer to Chapter 1 for soldering tips and techniques and for generic stripboard building guidelines.

Build the stripboard carefully following the circuit diagram shown in Figure 12-3. SW2 and SW3 are mounted off the board and are connected via three flying leads, and SW+ is a common connection that is connected to one leg of both switches. Connect the other leg of each switch to the SW2 and SW3 points accordingly, which are shown on the circuit diagram. You should end up with a board that looks like the one shown in Figure 12-4. Do not fit IC1 into the DIL socket at this stage.

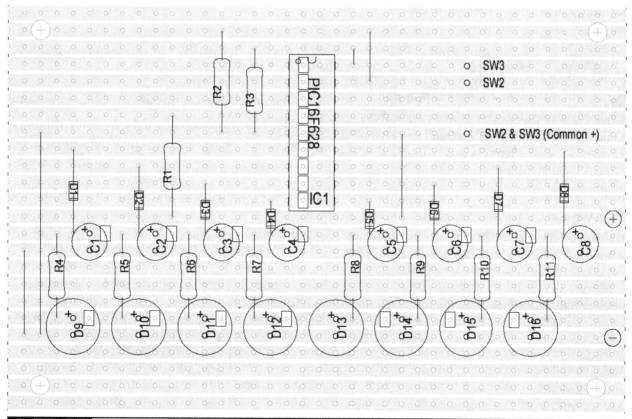

Figure 12-3 Stripboard layout for the LED scanner

132 Brilliant LED Projects

Figure 12-4 The completed stripboard

Before you program and fit IC1 to the stripboard, there are some basic tests that you can perform to make sure that the LEDs operate in the required manner; this also helps to cut down any fault-finding that may be required later. First of all, apply to the board a 4.5V supply using three AA batteries and make sure that this voltage is present at pin 5 (−) and pin 14 (+). There should also be +4.5V present at pin 4. If this is the case, then fit a short jumper wire between pin 14 and pin 6, as shown in Figure 12-5, and this should cause LED D9 to illuminate.

Now remove the link from pin 6, and the light from the LED should slowly be extinguished after a second or so. Repeat this method by keeping one end of the jumper wire connected to pin 14 and connecting the wire link to pins 6, 7, 8, 9, 10, 11, 12, and 13 one at a time in the same manner; this should prove the correct operation of each LED and its associated RC network in turn (see Figure 12-6).

Once you are satisfied that all of the LEDs work okay, you can remove the three AA batteries from the battery holder; it is now time to program IC1.

The PIC Microcontroller Program

If you were to unwrap a brand new microcontroller from its antistatic packaging and insert it into the DIL socket, you would be very disappointed because the circuit would not work. The reason is that you first need to write a software routine (the program) that will make the microcontroller interact with the circuitry around it and perform what you want it to do. This program also has to be converted into a suitable hex code and programmed into the microcontroller.

Chapter 12 ■ Introducing the PIC16F628-04/P Microcontroller: LED Scanner **133**

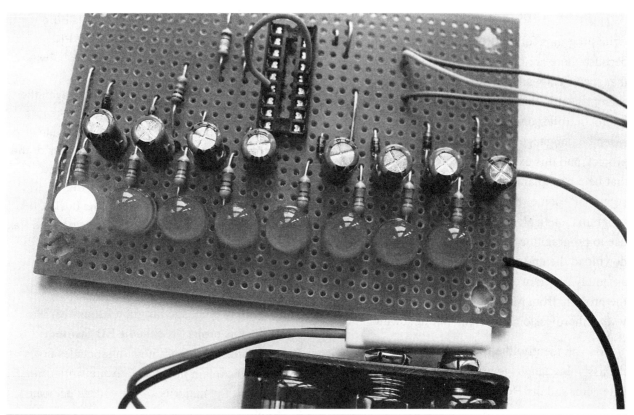

Figure 12-5 Connect a wire link between pin 14 and pin 6.

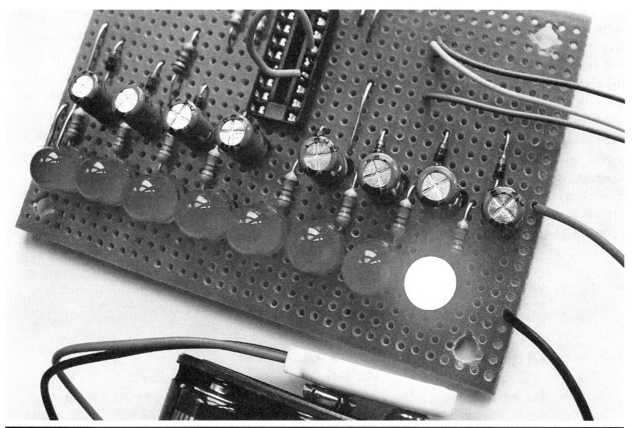

Figure 12-6 Test each of the LEDs in turn until you reach pin 13.

This book does not go into detail about how to write programs for the PIC Microcontroller because there are many excellent books available that specifically address how to do this. However, each project in this book that uses a microcontroller does include an extract of the assembly language program that I wrote for the project, and this extract also includes some notes that help to explain how the program works. I have converted each of these assembly language programs into a Hex file, and it is this file that you use to program the PIC Microcontroller. You can download the up-to-date versions of both the assembly program and the Hex files for each of the projects from McGraw-Hill's website at www.mhprofessional.com/computingdownload.

You can identify the file types easily because the assembly language program has a .asm extension and the Hex file has a .hex extension.

TIP If for some reason you are unable to download the software files from the website, you can manually type the hex code listings shown in each chapter into Notepad in Windows. After you finish typing the code into Notepad, select File | Save As. In the Save As dialog box, choose a location in which to store the file, make sure that Save As Type is set to All Files, give the file a name with a .HEX extension, and then click Save. (So, for example, you would save the Hex file for this project as **Led Scanner Project.hex**). This saved file will be recognized by the PIC programming software. The file has to be saved in Notepad because the data saved in these files is pure text, which is the form required by the PIC Microcontroller programming software.

CAUTION If you use this method, even the slightest typing error will cause the program to not run as expected; this could also mean that some of the input and output pins of the microcontroller are configured incorrectly and connected to a part of the circuit that they shouldn't be, potentially damaging the device. It is therefore recommended that you download the Hex files rather than typing them out manually.

The Assembly Program

The assembly language program listing that I wrote for this project is called **LED Scanner Project.asm**. The program listing includes notes that help to explain how the software works. You will see that the majority of the code is dedicated to monitoring the speed switches SW2 and SW3, which are connected to input Ports A0 and A1, and to creating the relevant delays to adjust the speed of the LED scanning. The "speed" variable is initially set to 30 upon power-up. This value is increased or decreased between the values of 1 and 60 whenever the speed switches are pressed. The eight I/O ports of Port B are configured as outputs, and a single output bit is made to roll from left to right and right to left along the outputs in a never-ending loop. These outputs are connected to the eight LEDs, and the "rolling bit" operation creates the LED scanning effect. An extract of the display routine is shown here:

```
LOOP1      movlw B'00000001'   ;move 1 to w
           movwf PORTB         ;send it to Port B to illuminate the first LED
           bcf STATUS,C        ;clear the carry flag

LOOP2      call PAUSE          ;create a delay
           call GETKEY         ;test to see if a key has been pressed
           rlf PORTB,F         ;rotate Port B output to move the LED to the left
           btfss STATUS,C      ;if carry flag is set then LED B7 is lit, skip a line
```

```
            goto LOOP2             ;loop around to continue moving the LED

            movlw B'10000000'      ;activate the LED on Port B,7
            movwf PORTB            ;
            bcf STATUS,C           ;clear the carry flag

LOOP3       call PAUSE
            call GETKEY
            rrf PORTB,F            ;rotate Port B right to move the LED the opposite way
            btfss STATUS,C         ;if carry flag is set then LED B0 is lit, skip a line
            goto LOOP3             ;loop around to continue moving the LED

            goto LOOP1             ;start the sequence all over again
```

The Hex File and Hex Code Listing

The LED Scanner Project.asm assembly program listing has been converted into a suitable Hex file to enable you to download it into the PIC Microcontroller. The following text shows what the Hex file looks like. This file is called **LED Scanner Project.hex** and can be downloaded from the McGraw-Hill website.

:020000000528D1
:08000800052807309F00850167
:10001000860183160330850000308600823081001F
:1000200083121E30A10001308600031024202D20F1
:10003000860D031C162880308600031024202D20F6
:10004000860C031C1E2813282108A0008B010B1D01
:1000500027280B11A00B272808000508003A0319D0
:1000600008000518A10A8518A10321083D3A0319C3
:10007000040282108003A031D08003C30A100080078
:060080000130A1000800A0
:02400E00303F41
:00000001FF

Program the PIC Microcontroller

If you already have your own preferred method of programming PIC Microcontrollers, simply download the Hex file for this project from the McGraw-Hill website to your PC and then use this file to program IC1 using your preferred hardware. If you are new to programming PIC Microcontrollers, then please read on to see how easy it is to do.

NOTE The following programming procedures have been tested using a PC running Windows XP; they have not been tested on other operating systems.

To program the chip, you need a dedicated programmer solution. I chose to use Microchip's PICkit 2 Development Programmer/Debugger and the programming software that is supplied with it. The PICkit 2 programmer contains a mini-USB connection, which is used to connect it to your PC, and a 6-pin output socket, which has the required outputs to program the PIC Microcontroller.

As you may recall from Chapter 1, the six output connections from the PICkit 2 programmer are configured in a specific order and need to be connected to the PIC16F628-04/P as shown in Chapter 1. The 6-pin connections of the PICkit 2 programmer, and how they should be connected to the PIC16F628-04/P microcontroller, are shown below (pin 1 of the programmer is denoted by the white triangle):

- Pin 1: $\overline{\text{MCLR}}$—connect to pin 4 of the PIC16F628-04/P
- Pin 2: Vdd Target (+V)—connect to pin 14 of the PIC16F628-04/P
- Pin 3: Ground (0V)—connect to pin 5 of the PIC16F628-04/P
- Pin 4: Data—connect to pin 13 of the PIC16F628-04/P
- Pin 5: Clock—connect to pin 12 of the PIC16F628-04/P
- Pin 6: Aux—not used

Pins 1, 2, 3, 4, and 5 of the PICkit 2 programmer need to be connected to the PIC Microcontroller to program it. I built on breadboard a programming interface that allows IC1 to be programmed in isolation from the LED scanner circuit. More details of how to build this programming interface for yourself are outlined in Chapter 1.

Figure 12-7 shows the PICkit 2 programmer plugged into the SIL pin header (which I fitted to the breadboard), ready to program the PIC Microcontroller. Note that pin 1 of the PIC Microcontroller is positioned toward the top of the breadboard in Figure 12-7. After you connect the programmer to the breadboard, you can program the hex code into the PIC Microcontroller via the programming software, as explained a bit later in the chapter. The programmer provides the required power and programming voltages to program the microcontroller.

Once you have installed the PICkit 2 programmer, and the programming software (that comes with the PICkit 2 programmer) onto the hard drive of your PC, by following the instructions supplied with it, you should then download the Hex file for this project, **LED Scanner Project.hex**, and save it in a folder on your PC.

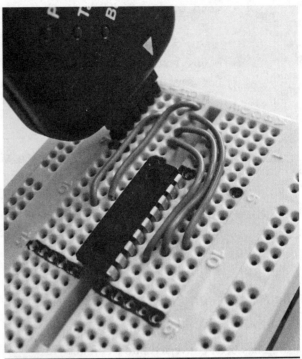

Figure 12-7 The PIC Microcontroller connected to the breadboard interface

Ensuring that your PC is booted up first, the steps required to program IC1 are as follows:

1. Connect the PICkit 2 programmer to your PC using a suitable USB cable.

2. Insert IC1 into the breadboard interface and then plug the PICkit 2 programmer into the pin header on the breadboard, as shown in Figure 12-7.

3. Open the programming software on your PC. It should automatically find the PICkit 2 programmer and recognize that it is connected to a PIC16F628 device.

4. Set the VDD PICkit 2 voltage setting in the software to 4.5 volts, as shown in Figure 12-8. Leave unchecked the On and /MCLR check boxes that are next to the voltage setting.

5. Click the Erase button to wipe any data that may already be in the microcontroller, and wait for the software to confirm that this has been completed, as shown in Figure 12-8.

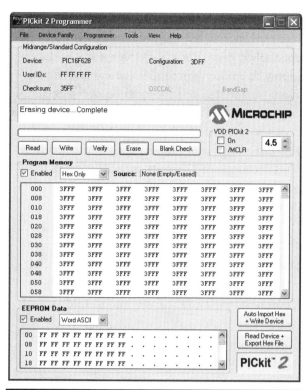

Figure 12-8 Set the programming voltages in the software before clicking Erase.

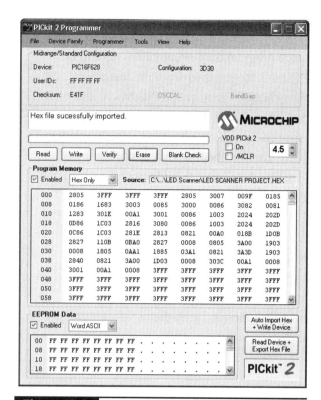

Figure 12-9 Import the Hex file.

6. Choose File | Import Hex, find the **LED Scanner Project.hex** file for this project on your PC, and import the file into the software, which should display the message shown in Figure 12-9 for a successful import.

7. Make sure that the VDD PICkit 2 setting for the programmer is still set to 4.5 volts, and then click the Write button. This should automatically transfer the Hex file into the microcontroller. If programming has been successful, then after a few seconds the software will tell you so, as shown in Figure 12-10. If you encounter a problem downloading the Hex file into the microcontroller using the PICkit 2 programmer, consult the software's very helpful user guide, which can you access via the Help menu; this guide should help you to find the problem.

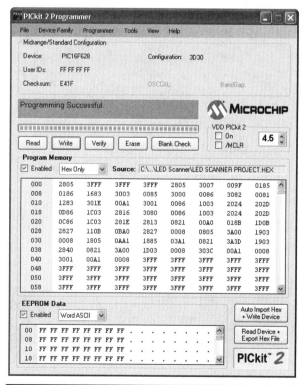

Figure 12-10 The Hex file has been transferred to the microcontroller successfully.

See the Scanner in Action

Once programming is complete, remove the PICkit 2 programmer from the breadboard and then carefully remove the PIC Microcontroller and insert it into the DIL socket of the project stripboard, making sure it is correctly oriented. Next, connect the three AA batteries to the stripboard, and you should immediately see the LED scanning effect in action, as shown in Figure 12-11. You can increase and decrease the LED scanning speed by pressing SW2 and SW3, respectively. If the project does not work as expected, then you will need to double-check your stripboard layout and also make sure that IC1 has been programmed with the correct Hex file.

NOTE The microcontroller circuit layout for this project doesn't contain a decoupling capacitor across the + and − battery supply to smooth the supply voltage and to help avoid potential spurious triggering of the circuit. I didn't include a decoupling capacitor because the circuit worked well without one. If you experience a problem with your project, try fitting a 100nF or 0.1µF (minimum 10V rated) capacitor across the positive and ground rails of the circuit to see if that helps to alleviate the issue.

Now all you need to do is find a suitable housing for the stripboard and mount the speed switches in a convenient location. You could incorporate the LED scanner into a child's toy, or use it as an eye-catching badge that you could wear on your clothing.

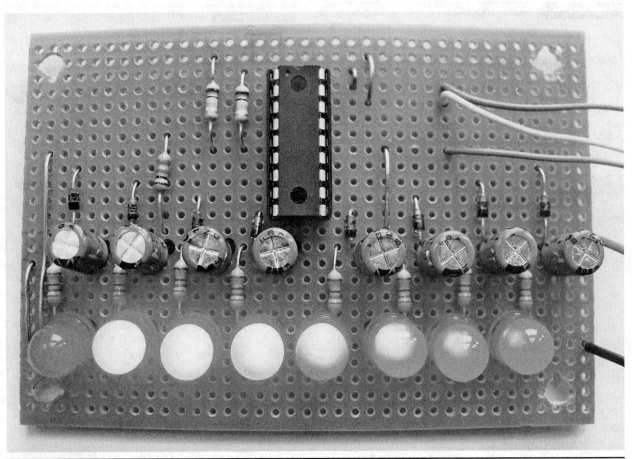

Figure 12-11 The LEDs scan from side to side.

CHAPTER 13

LED Light Sword

How would you like to make your own compact LED light sword to fend off unwanted enemies? The project in this chapter shows you how to build an LED light sword that utilizes a PIC Microcontroller to create a slightly different sequencer effect. The effect generated when you activate the sword is a rising column of light, which starts at the bottom of the sword and then rises to the top. The software code that you program into the microcontroller also strobes the eight LEDs very quickly to allow a special "clash" facility. The instructions in this chapter take you through each of the constructional steps required, and show you how to use some salvaged plastic tubing to create a finished sword that looks like the one shown in Figure 13-1.

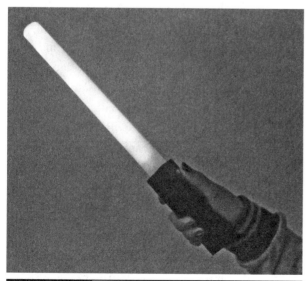

Figure 13-1 LED light sword

Project 12
The LED Light Sword

If you have not read through Chapter 12, yet it is recommended that you do so to understand some of the key features of the PIC16F628-04/P Microcontroller, and to also understand how to program this device. The project specifications for the LED light sword are outlined below.

PROJECT SPECIFICATIONS

- The LED light sword incorporates eight large 10mm red LEDs.
- The visual effect is a rising column of light that emanates from the handle to the tip of the sword.
- The circuit is driven by the PIC16F628-04/P Microcontroller.
- A "power-up" push button creates the visual light effect.
- A "clash" push button brightens the light output to simulate swords clashing together.
- The supply voltage is 4.5 volts.

How the Circuit Works

The circuit diagram for the LED light sword is shown in Figure 13-2. As mentioned at the beginning of the chapter I have utilized a PIC

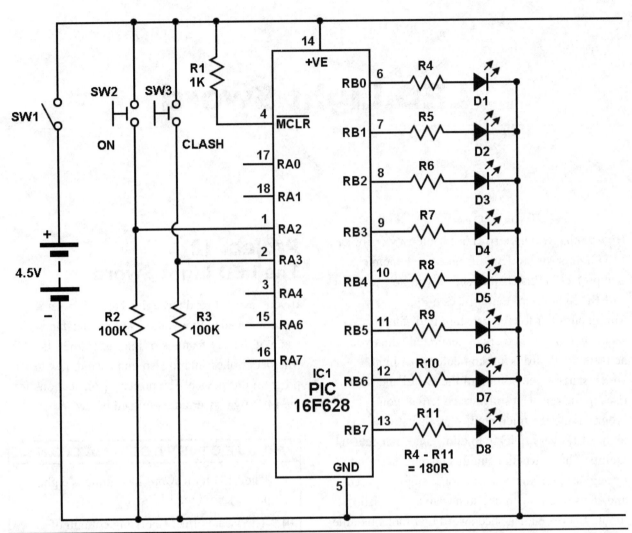

Figure 13-2 The circuit diagram for the LED light sword

Microcontroller at the heart of the circuit, both to simplify the layout and to produce a compact stripboard that will fit within the handle of the sword. I initially considered using an LM3914 bar-graph driver chip in bar mode to create the visual effect required for this project, but I ultimately decided that by using a microcontroller, I could simplify the circuit layout and incorporate some additional features. (You'll have a chance to use the LM3914 bar-graph IC if you choose to build the digital oscilloscope project in Chapter 17.)

The circuit layout is fairly simple and is very similar to that of the LED scanner project in Chapter 12. The circuit is powered by a 4.5V supply in the form of three AAA batteries. This activates the microcontroller (IC1) and its program upon power-up. SW1 is the power switch, and this is optional. If you decide not to incorporate SW1, you will need to bear in mind that in quiescent mode (that is, when no LEDs are illuminated) the circuit draws just under 1mA, which is fairly low, but the batteries would probably last only a few weeks before wearing out. It is up to you to decide if you want to include SW1; the design shown here doesn't include it because the enclosure that I used has limited space.

SW2 is the "power-up" push button, which is connected to pin RA2 that is configured as an input. Normally this input is kept low via resistor R2 until SW2 is pressed. When SW2 is pressed, the software recognizes a high signal on input RA2 and activates the LED illumination sequence, which activates LEDs D1 to D8 one at a time starting at D1 and ending at D8. D1 to D8 are connected to Ports B0 to B7, which are configured as outputs and are driven via series resistors R4 to R11. You need to keep SW2 pressed to keep the LEDs illuminated. At this point the software causes the eight LEDs to produce a strobe effect by switching them all on and off very quickly, fast enough to make it look as though all of the LEDs are illuminated but at a slightly reduced brightness. This helps to reduce the current consumption of the circuit because the eight LEDs are only lit for half of the time, although this is not apparent to the human eye. When you release SW2, the column of light descends back from D8 to D1 until all of the LEDs are turned off again.

Push button switch SW3 is an optional feature and is called the "clash" button. This switch is connected to input RA3 and is recognized by the software only after SW2 is pressed and all of the LEDs are illuminated. If SW2 and SW3 are pressed at the same time, the software no longer causes the eight LEDs to strobe; instead, it switches them all on permanently, which has the effect of brightening the LED light sword. This feature could be used to emulate the sword clashing with another object or sword. You may decide not to include SW3 at all, which is fine—the circuit will work without it installed. Alternatively you may also decide to link out SW3 from the positive (+ve) supply rail to input pin RA3 to make the sword look brighter all of the time. This would bypass the strobe effect in the software, making the LEDs look brighter, but it would also increase the current consumption of the sword.

Parts List

NOTE The Supplier and Part Number column of the following table lists specific parts that I used in this project. Refer to the appendix for additional details about acquiring your parts.

The parts you'll need for the LED light sword project are listed next.

		PARTS LIST	
Code	Quantity	Description	Supplier and Part Number
IC1	1	PIC16F628-04/P Microcontroller	RS Components 379-2869 (Mfr: MicrochipTechnology Inc. PIC16F628-04/P)
R1	1	1KΩ 0.5W ±5% tolerance carbon film resistor	—
R2, R3	2	100KΩ 0.5W ±5% tolerance carbon film resistor	—
R4–R11*	8	180Ω 0.5W ±5% tolerance carbon film resistor	—
D1–D8	8	10mm red LED V_F (typical) = 1.8V, I_F (typical) = 20mA	RS Components 577-768
SW1	1	Single-pole panel mount toggle switch, 2A rated (optional; see text)	RS Components 710-9674
SW2, SW3	2	Single-pole normally open panel mount switch (100mA)	RS Components 133-6502

(continued)

PARTS LIST (continued)

Code	Quantity	Description	Supplier and Part Number
Hardware	1	Stripboard, 0.1" (2.54mm) hole pitch, 25 holes wide by 9 tracks high	—
Hardware	1	18-pin DIL socket	—
Hardware	1	12-pin SIL header pin strip	—
Hardware	1	AAA battery holder (four AAA batteries; see text)	RS Components 512-3568
Hardware	3	AAA battery (1.5V)	—
Hardware	1	Small, narrow enclosure, approx. 4.9" (124mm) long × 1.3" (33mm) wide × 1.2" (30mm) deep	Maplin FT31
Hardware	1	1.06" (27mm)-diameter acrylic tube or similar (see text)	—
Hardware	—	Flexible interconnecting wire, solid tinned copper wire, cable ties, cable tie base, and tracing paper	—

* Note: If you use LEDs that have different V_F and I_F values to those that are in the parts list, then you may need to alter the resistance and wattage values of these LED series resistors. Please refer to Chapter 2, which explains how to do this, and also refer to Chapter 12, which outlines the current capabilities of the PIC16F628-04/P microcontroller. I used a value of approximately 15mA for I_F in my calculations for this project.

How to Make the Enclosure

Take another look at the finished LED light sword shown in Figure 13-1 (at the beginning of the chapter) and you will see that there are two parts to the enclosure, the handle and the light tube.

The Handle

The handle that I used is a small, narrow enclosure. It is just the right size to serve as a handle to grip and is also large enough to incorporate the battery holder, the stripboard, and the two switches. When I initially designed this project I started to build my sword by carefully drilling three holes in the handle enclosure, as shown in Figure 13-3, which are required for the two switches and the light tube. You need to ensure that these holes are large enough to accommodate these three parts. I positioned the center of the hole for SW2 approximately 0.8 inch (21mm) from the light tube end of the enclosure, and positioned the center of the hole for SW3 approximately 1.2 inches (31mm) from the light tube end of the enclosure. This configuration makes both switches easy to operate using my thumb and index finger when holding the handle.

Figure 13-3 The predrilled enclosure

TIP Solder the electronic components to the stripboard first before drilling the two switch holes in the enclosure. You can then use the finished stripboard as a template to mark out the switch positions on the enclosure. Because there is little space inside the enclosure, this will help to ensure that you can install the two switches in a position where they do not come in contact with any of the electronic components.

The Light Tube

You could use a piece of clear acrylic tube to create the light tube. However, as you probably have realized by now, I like to utilize salvaged parts to build my projects, and this project is no exception. I decided to salvage a clear tube that used to be part of a bubble-maker toy like the one shown in Figure 13-4.

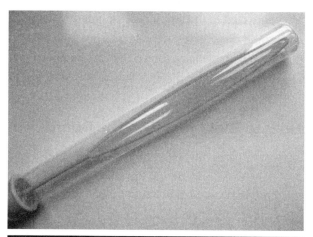

Figure 13-4 I salvaged the light tube from a bubble-maker toy like this one.

The bubble maker from which I salvaged the tube is the type that you can purchase in a toy shop or online. The clear tube is normally filled with a soapy liquid. The bubble maker has a handle with a loop so that you can dip the loop into the liquid and then create big, soapy bubbles by waving the loop in the air. I used the liquid container, shown in Figure 13-5, as my light tube. (Unfortunately, I couldn't use the handle and loop in this project. Of course, I'll save it for possible use in a later

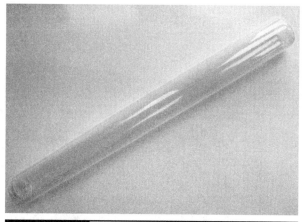

Figure 13-5 The empty light tube

project.) I lightly sanded the outer face of the tube using fine-grade sandpaper, which helps to diffuse the light output from the LEDs. The tube that I used is approximately 10.5 inches long and its diameter is just over 1 inch, which makes my LED light sword quite compact. You might decide to use a longer tube for your project, but make sure the diameter is sufficient to feed the string of LEDs into it.

Conveniently, the bubble tube that I used has a threaded end, which I screwed into the hole that I had drilled into the top of my handle enclosure. I then glued the tube into place using an epoxy resin to provide a strong, secure fixing. Next, I ensured that all of the components would fit neatly inside the handle enclosure, as shown in Figure 13-6. Finally, I removed the battery enclosure, the stripboard, and the switches and prepared to build the electronics. If you are unable to get hold of a threaded tube you may need to use a larger enclosure for the handle, and you will need to drill a larger hole to accommodate the full diameter of the tube. If you use this method you would need to ensure that you use plenty of epoxy resin, and probably a screw and nut to hold the tube in place. Alternatively, you could use a wider diameter tube and find a way of incorporating all of the electronics inside the tube; this would remove the need for a separate handle enclosure altogether.

144 Brilliant LED Projects

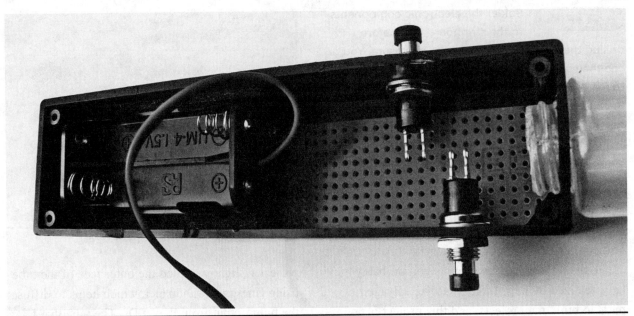

Figure 13-6 Make sure that everything fits into place.

Stripboard Layout

The stripboard layout for the LED light sword project is shown in Figure 13-7. The IC control part of the circuit is built on a small piece of stripboard that is only 25 holes wide by 9 tracks high, which fits neatly inside the narrow enclosure. Note that you need to make 17 track cuts before you start to solder the components in place (shown as white rectangular blocks), including those underneath the IC's DIL socket and underneath each of the series resistors, R4 through R11.

The eight LEDs are mounted remotely from the board, in the light tube part of the sword. The LEDs are connected to the stripboard via interconnecting wires, and these wires are soldered onto the SIL header pins, which are shown on the stripboard layout; more on this shortly.

How to Build and Test the Board

NOTE Please refer to Chapter 1 for soldering tips and techniques and for generic stripboard building guidelines.

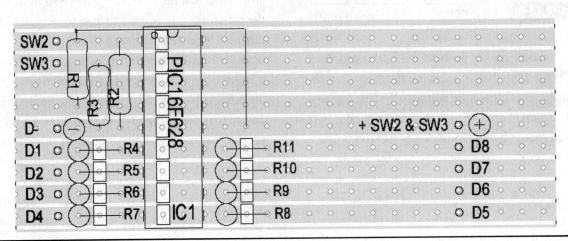

Figure 13-7 Stripboard layout for the LED light sword

Figure 13-8 The completed stripboard layout

Build the stripboard layout by closely following the diagram shown in Figure 13-7. Also notice that resistor R1 reaches around the top of IC1. Once complete, you should end up with a layout that looks like the one shown in Figure 13-8.

After you have built the stripboard, you may decide to test the circuit before you insert IC1, in a similar manner to the testing described in previous projects. For example, you could apply power to the board and make sure that the voltages are as expected at pins 4, 5, and 14. You could then apply a + voltage to pins 6–13 one at a time in turn to ensure that the voltages are as expected for each of the relevant SIL pins.

How to Make the LED String

The eight LEDs need to be wired and soldered in a row or a string that will fit neatly inside the light tube part of the enclosure. First of all, measure the length of the light tube that you are using and divide this length by 8 so that you know how far apart to wire each LED. Then, to form a framework that you can solder each of the LEDs to and that will also strengthen the string of LEDs, find a fairly solid length of tinned copper wire to use as the common cathode (−) connection. Cut the copper wire to the same length as the clear tube of the sword. Solder the end of the copper wire to a piece of flexible wire, which you will eventually solder to the − SIL pin on the stripboard. Figure 13-9 shows how to solder the LEDs together.

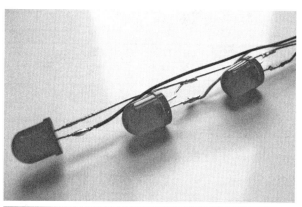

Figure 13-9 Solder the LED string together.

Solder the anode connection (+) of each LED to a length of flexible wire that is long enough to reach the SIL pins on the stripboard. To make it easy to identify each of the wires, use a different color of cable for each, or label each wire with a piece of masking tape with the relevant LED number on it. Once you have soldered all eight LEDs to the copper wire, you should end up with a string of eight LEDs connected to nine cables (of different colors or labeled with masking tape), which you will soon solder to the stripboard. Before you insert the LED string into the enclosure, you might want to test that each of the LEDs illuminates correctly by using a 3V battery and an 180Ω resistor. Once you are happy with the LED string, you can carefully feed it into the clear tube through the handle enclosure, as shown in Figure 13-10. You might want to apply a small blob of glue to the top LED to secure it to the top of the tube.

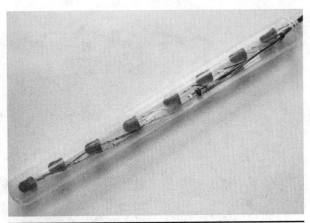

Figure 13-11 The LED string in place

Putting It All Together

Now that you have built all the elements of the LED light sword, completing it is just a simple exercise of putting them all together. The first step is to insert the stripboard layout into the handle enclosure and solder the LED wires to the relevant SIL pins. Cut the wires to size so that they reach the SIL pins and still have a little spare just in case you need to resolder them in the future.

The next step is to fit the two push button switches, SW2 and SW3, making sure that they do not touch any of the components on the stripboard, and then solder them to the relevant SIL pins. Ensure that the two switches are soldered together and linked to the + battery pin.

Finally, insert the AAA battery holder into the handle and solder the + and − cables to the relevant SIL pins. Although the battery holder accepts four AAA batteries, you can't use four batteries because that would produce an output voltage of over 6 volts, which is too high for IC1. Thus, you need to carefully solder a wire link across one of the battery terminals (refer to "How to Build and Test the Board" in Chapter 11 for an explanation of how to do this), making sure that you do not melt the plastic enclosure. That should complete the electronics construction, and the stripboard layout should look like the image shown in Figure 13-12.

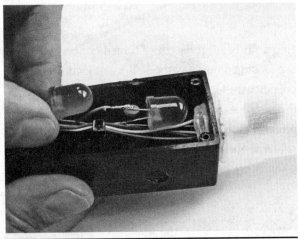

Figure 13-10 Feed the LED display into the light tube.

> **TIP** Wrap some electrical insulating tape around the soldered joints to prevent the LED leads from shorting out on each other.

The LED string should end up looking similar to Figure 13-11 (I took this image before I sanded the clear tube and glued it to the handle so that you can see what the string looks like when it is in place).

Secure the nine flying LED wires into place by using a cable tie and cable tie base, which you then fit inside the handle enclosure.

You can make the battery holder fit snugly in the enclosure by applying sponge adhesive tape to the inside of the enclosure. The completed handle enclosure should look like Figure 13-13.

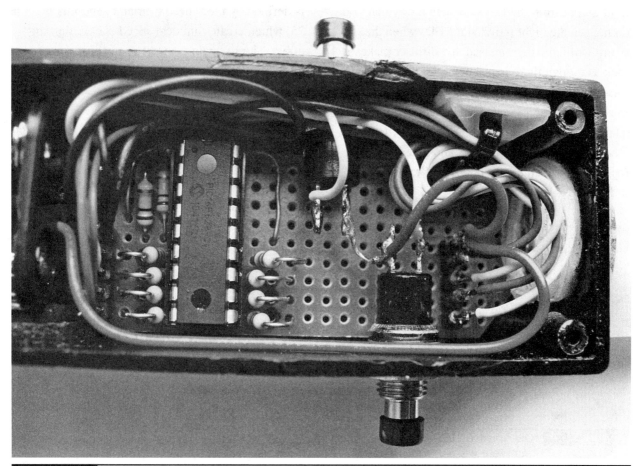

Figure 13-12 The enclosure should now look something like this.

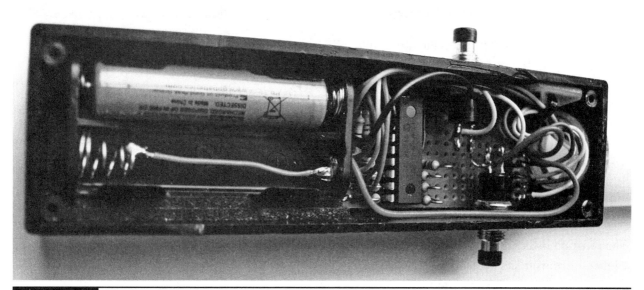

Figure 13-13 The handle enclosure with the battery holder and stripboard

One finishing touch is required, and that is to make a paper diffuser sleeve that covers the light tube. The purpose of this sleeve is to diffuse the light output from the LEDs in such a way that you do not see the eight individual LEDs when the sword is illuminated; instead, the diffuser makes it look as though there is just a single shaft of light. I used a piece of A4-size tracing paper, which I cut to size, wrapped around the tube, and then secured using double-sided adhesive tape. Finally, I cut a circular piece of tracing paper and glued it to the top of the paper tube, producing the result shown in Figure 13-14.

Figure 13-14 The finished LED light sword, complete with its light diffuser sleeve

The PIC Microcontroller Program

Now that you've built the LED light sword, you need to program IC1 and install it on the stripboard. You can download the assembly program and Hex file from the McGraw-Hill website, at www.mhprofessional.com/computingdownload. You can then program IC1 with the Hex file as explained in Chapter 12.

The Assembly Program

The program listing for this project is called **LED Light Sword.asm** and contains plenty of notes to explain how it works. You will notice that this program has two look-up tables that define the sequence of LEDs on power-up and power-down. The speed at which the LED light shaft illuminates is defined by the "speed" variable, and this is set at 25, which creates an ideal speed for the lighting effect. The following extract of the program shows the look-up tables and how each byte illuminates the LED sword sequence in turn:

```
POWERON:   addwf PCL,F          ;LED sequence
                                ;for power up
           retlw b'00000000'    ;all LEDs off
           retlw b'00000001'    ;first LED on
           retlw b'00000011'    ;2 LEDs on
           retlw b'00000111'    ;3 LEDs on
           retlw b'00001111'    ;4 LEDs on
           retlw b'00011111'    ;5 LEDs on
           retlw b'00111111'    ;6 LEDs on
           retlw b'01111111'    ;7 LEDs on
           retlw b'11111111'    ;8 LEDs on

POWEROFF:  addwf PCL,F          ;LED sequence
                                ;for power
                                ;down
           retlw b'11111111'    ;8 LEDs on
           retlw b'01111111'    ;7 LEDs on
           retlw b'00111111'    ;6 LEDs on
           retlw b'00011111'    ;5 LEDs on
           retlw b'00001111'    ;4 LEDs on
           retlw b'00000111'    ;3 LEDs on
           retlw b'00000011'    ;2 LEDs on
           retlw b'00000001'    ;1 LED on
           retlw b'00000000'    ;all LEDs off
```

The Hex File

The Hex file that you need to transfer into IC1 is called **LED Light Sword.hex** and is shown below. Program IC1 by using this file, as outlined in Chapter 12, or by using your own preferred programming method.

```
:020000001928BD
:080008001928820700340134BD
:10001000033407340F341F343F347F34FF348207F6
:10002000FF347F343F341F340F340734033401343A
```

```
:10003000003407309F008501860183160C3085004F
:100040000030860081308100831219308A2008601C1
:10005000051D2828A0012008052086004920A00AA7
:100060002008093A031935282B28A001051D3F282F
:10007000FF308600851936280030860036282008 93
:100080000F2086004920A00A2008093A03192828D1
:100090003F282208A1008B010B1D4C280B11A10B3E
:0400A0004C280800E0
:02400E00303F41
:00000001FF
```

The Final Tests

Once you have programmed IC1, fit it into the DIL socket on the stripboard and install the three AAA batteries into the battery holder. When you apply power to the circuit, nothing should happen. Press and hold the power-up button (SW2) using your thumb, and a shaft of LED light should start to rise from the bottom of the sword to the top. As soon as the light reaches the top of the sword, the light should dim slightly as the software causes the LEDs to produce a strobe effect. Now keep your thumb on SW2 and simultaneously use your index finger to press the clash button (SW3), and the light shaft should brighten up; releasing your finger from SW3 causes the LEDs to revert to the strobe effect and produces a reduced light output. Remove your thumb from the power-up button, and the shaft of light should drop down to the bottom of the sword until all the LEDs are off again. If you find that the LED light sword doesn't operate in this way, then remove the batteries immediately and check the board over for constructional errors or for any stray solder that might be shorting out the copper tracks. You may also want to try re-programming IC1 to make sure that the hex code has been installed correctly.

 The microcontroller circuit layout for this project doesn't contain a decoupling capacitor across the + and − battery supply to smooth the supply voltage and to help avoid potential spurious triggering of the circuit. I didn't include a decoupling capacitor because the circuit worked well without one. If you experience a problem with your project, try fitting a 100nF or 0.1μF (minimum 10V rated) capacitor across the positive and ground rails of the circuit to see if that helps to alleviate the issue.

Time to Play

Now it is time to play. Like most of the projects in this book, the lighting effect is best appreciated in the dark. If you decide to build two LED light swords so that you can have a duel, you need to make sure that you don't smash the swords against each other because this will probably damage the enclosure, and possibly the electronics. Miming a fight is probably the best thing to do. If you fancy making a larger sword, you might consider modifying the circuit and software design so that more than eight LEDs are used, but don't forget the output current capabilities of the PIC16F628-04/P device. You may also need to incorporate additional driver circuitry, and this may mean using a larger piece of stripboard and a larger enclosure to house the electronics.

Finally, you've probably realized that there is no sound effect circuitry incorporated into this design, so, like me, you will need to use your voice and make your own sound effects as you wave the sword through the air!

CHAPTER 14

A Manually Operated Sequencer: Invisible Secret Code Display

THE HANDHELD DEVICE DESCRIBED IN THIS CHAPTER'S project allows you to transmit to a friend ten individual coded shapes using a display composed of seven LEDs in a star formation. The key feature of this device is that you can't see its coded image unless you know how to view it. This is because the display transmits the image using infrared light, which can't be seen by the human eye. The image can be seen only by viewing it through a digital camera.

Any budding secret agent worth their salt should have one of these devices. Even if your fellow spy is across the room from you, you could still "see" the coded image by using your digital camera to zoom in on their handheld transmitter. The handheld transmitter that I built is shown in Figure 14-1. With a bit of creativity, you could also use this device as part of a magic trick.

Project 13
The Invisible Secret Code Display

This project utilizes seven infrared (IR) LEDs to create the invisible effect described in the chapter introduction. Again, the light output from this type of LED can't be seen by the human eye. IR LEDs are used in many household remote controls, such as those that you use to change channels on your TV or to control your DVD player. Your remote control pulses the LED in a particular sequence, and this signal is monitored by an infrared receiver in your TV set and decoded accordingly. You can see the light output from these LEDs by viewing them through a digital camera because the image sensors used in modern digital cameras are able to "see" infrared light. You can confirm this concept by pressing any button on your TV remote and viewing the LED transmitter in your remote control through a digital camera.

This project utilizes this principle by using seven infrared LEDs laid out in a particular pattern. Ten different patterns can be generated on the display by adjusting a coding switch that is mounted on the back of the device. The ten different shapes represent the numbers 0 to 9 and

Figure 14-1 The invisible secret code display

151

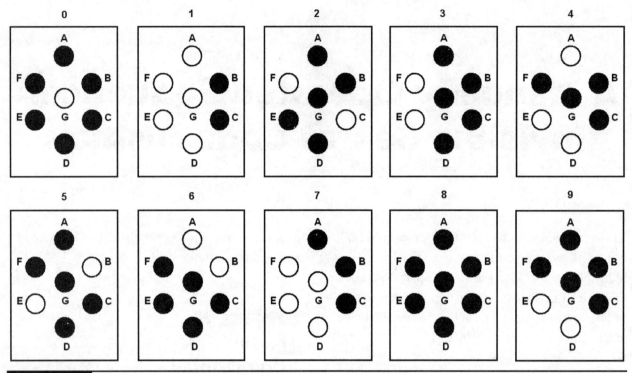

Figure 14-2 The ten different image codes that can be transmitted

are depicted in Figure 14-2. The LED positions are laid out in the same format as a seven-segment display, so if you have a look at Figure 7-3 in Chapter 7 you should be able to recognize the number patterns generated.

PROJECT SPECIFICATIONS

- The handheld unit is discreet and includes an encoder switch and infrared display.
- A ten-position encoder switch generates ten different images.
- A "transmit" push button activates the infrared display.
- The supply voltage is a low 3 volts.

How the Circuit Works

The circuit diagram for the invisible secret code display is shown in Figure 14-3. The circuit uses a single IC to create the codes, a 4511B BCD to seven-segment driver. This CMOS device reads a binary coded decimal (BCD) signal on its inputs and converts it to an output code that normally drives a seven-segment display. Usually this BCD signal is generated by another digital IC, but in this project you will manually control the BCD input by using a special BCD encoder switch. This is a ten-position switch that provides the four digital outputs required to control the 4511B.

Resistors R1 to R4 hold the inputs of IC1 low, and these inputs are switched high in a BCD format whenever the ten-position switch is adjusted. The 4511B device has internal driver circuitry that allows it to drive LEDs (normally a seven-segment display), and so is able to drive the seven IR LEDs (D1 to D7) in this project. The series resistors (R5–R11) are still required to limit the current to each LED, and these are set to generate enough IR light output to be visible in the dark when viewed through your digital camera. The 4511B device also has some other features

Chapter 14 ■ A Manually Operated Sequencer: Invisible Secret Code Display

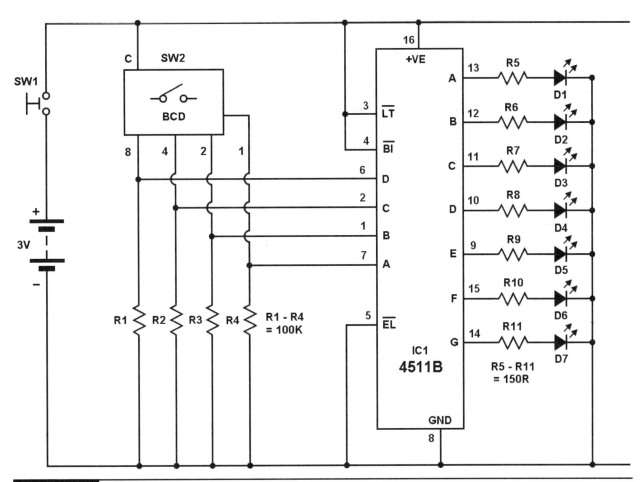

Figure 14-3 The circuit diagram for the invisible secret code display

that won't be used in this project, such as lamp test, blanking, and latch enable. These features are disabled by pins 3 and 4 being held high and pin 5 being held low.

IR LEDs normally have a lower forward voltage drop than some of their counterparts that produce visible light output. The type of LED used in this project has a V_F value of 1.3V, which means that a low voltage supply can be used. A 3V supply is used in this circuit. Although this is the lower voltage range that can be used by IC1, it seems to work well in this application; in fact, I have managed to get the device to work at voltages as low as 2.5V.

TIP The IR LED type that you need to use in this project has a tinted color to its lens and has no visible light emanating from it when it is activated. Some IR LEDs have a faint visible light output when activated; these are not suitable for this project.

SW1 is a momentary push-to-make switch that is used as a transmit button. When this button is pressed, the circuit is activated and the IR display shows the relevant code depending on the position in which SW2 is set. When the push button is released, the unit switches off, thus saving battery power when it is not in use.

Parts List

> **NOTE** The Supplier and Part Number column of the following table lists specific parts that I used in this project. Refer to the appendix for additional details about acquiring your parts.

The parts you'll need for the invisible secret code display project are listed in the table below.

Stripboard Layout

The stripboard layout for the project is shown in Figure 14-4. First, cut the stripboard so that it is 21 holes wide by 24 tracks high. I trimmed down a piece of stripboard that was originally 37 holes wide by 24 tracks high. You then need to make 20 track cuts in total; these are shown as white rectangular blocks. You also need to make two holes so that you can mount the board onto the internal mounting posts of the enclosure. If you use a different enclosure from the one identified in the parts list, then you may need to adjust the positions of these holes accordingly. It should be noted at this point that the BCD switch (SW2) is going to be soldered to the trackside of the board.

		PARTS LIST	
Code	Quantity	Description	Supplier and Part Number
IC1	1	4511B BCD to seven-segment decoder	RS Components 306-718 (or similar)
R1–R4	4	100KΩ 0.5W ±5% tolerance carbon film resistor	—
R5–R11*	7	150Ω 0.5W ±5% tolerance carbon film resistor (see text)	—
D1–D7	7	5mm infrared emitter LED (LD274) V_F (typical) = 1.3V, I_F = 20mA (typical)	ESR Electronic Components 720-530 or RS Components 654-8160
SW1	1	Single-pole normally open panel mount switch (100mA)	RS Components 133-6502
SW2	1	BCD ten-position rotary encoder switch	RS Components 708-3217 (Mfr: Knitter-Switch DRS61010)
Hardware	1	Stripboard, 0.1" (2.54mm) hole pitch, 37 holes wide by 24 tracks high (see text)	—
Hardware	1	16-pin DIL socket	—
Hardware	1	AAA battery holder (two AAA batteries)	RS Components 512-3552
Hardware	2	AAA battery (1.5V)	—
Hardware	1	Enclosure with integral battery compartment, approx. 4.1" × 2.4" × 1.1" (105mm × 61mm × 28mm)	RS Components 244-8555
Hardware	—	Double-sided adhesive strips, cable ties, and cable tie base	—
* Note: If you use LEDs that have different V_F and I_F values to those that are in the parts list, then you may need to alter the resistance and wattage values of these LED series resistors. Please refer to Chapter 2, which explains how to do this, and also read the "Future Modifications" section at the end of this chapter which discusses the current capabilities of IC1.			

Chapter 14 ■ A Manually Operated Sequencer: Invisible Secret Code Display 155

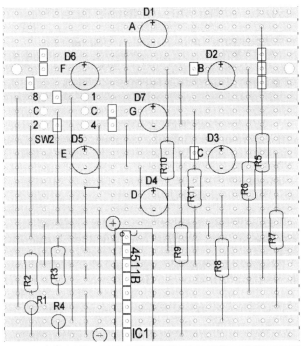

Figure 14-4 The stripboard layout for the invisible secret code display

Before you start to populate the board, create marks on it indicating the center points where each of the seven LEDs and the rotary shaft of SW2 will sit, and then overlay the board with a piece of tracing paper and transfer these markings onto the paper, as shown in Figure 14-5. You can then use this tracing paper as a template to mark on the front of the enclosure the hole positions of the seven LEDs and to mark on the back of the enclosure the hole position of the BCD switch; this is explained shortly.

Preparing the Enclosure

The enclosure that I used is a compact hand held case that includes an integral battery compartment. The case splits into two halves and comprises a lid, which the LEDs will protrude through, and the back of the enclosure, which will soon allow access to the BCD switch (SW2). Using your tracing paper template, mark the drilling positions for the seven LEDs onto the lid of the enclosure. Once the positions are marked onto the lid, drill the seven holes using a .25" (6mm) drill bit so that the holes are large enough for the IR LEDs to protrude through once the unit is complete. This process can be seen in Figure 14-6.

Figure 14-6 Use the tracing paper as a template before drilling the seven LED holes into the lid.

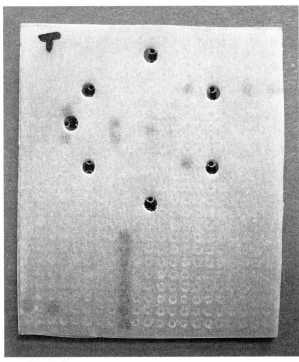

Figure 14-5 Mark the eight hole positions on the tracing paper.

Once the seven holes have been drilled into the lid, you may decide at this point to drill another hole into the side of the enclosure lid to accommodate the panel mount switch (SW1). This hole should be drilled in a similar position to the one in my enclosure lid, which is shown in Figure 14-7; this is so that once the switch is in place it does not come into contact with any of the electronics on the stripboard. Alternatively, you may prefer to drill this hole after you have built the stripboard, so that you are confident that the switch position can accommodate the electronics.

Now you should use the tracing paper template to identify the position of the BCD switch (SW2) and mark this out on the back of the enclosure, not forgetting that the BCD switch will be soldered to the track side of the board. Drill a small pilot hole in the back of the enclosure; this will be soon made larger to accept the BCD switch. Look ahead to Figures 14-9 and 14-12 to see where the BCD switch will be positioned on the back of the enclosure.

How to Build and Test the Board

NOTE Please refer to Chapter 1 for soldering tips and techniques and for generic stripboard building guidelines.

After you have prepared the enclosure, solder all of the components onto the stripboard (with the exception of the seven LEDs, for now) by following the stripboard layout shown in Figure 14-4. Before you solder the BCD rotary switch (SW2) to the track side of the board, carefully bend the six pins flat. The completed track side of the board is shown in Figure 14-8, which also shows the position of the BCD switch.

Place the stripboard inside the rear of the enclosure, as though you are going to screw it into place. This won't be possible yet because you need to increase the pilot hole size that you made for the BCD switch on the back of the enclosure. Use the board to identify and mark out where you need to

Figure 14-7 The lid of the enclosure has been drilled, including a hole in the side for SW1.

Chapter 14 ■ A Manually Operated Sequencer: Invisible Secret Code Display

Figure 14-8 Solder the BCD switch to the track side of the stripboard.

increase the hole position and then remove the board from the enclosure. Now with some careful measuring and drilling you should be able to make the hole large enough so that the BCD switch protrudes through the rear of the enclosure, as shown in Figure 14-9.

Figure 14-9 Make sure that the BCD switch protrudes through the back of the enclosure.

Now that the stripboard sits neatly inside the enclosure, you can measure the exact length required for the LED leads so that the LEDs protrude slightly through the holes in the enclosure lid once it is in place. Cut the LED leads to size and solder them onto the stripboard. The LEDs should stand up off the board, as shown in Figure 14-10, which gives you an idea of what the top of the completed stripboard should look like.

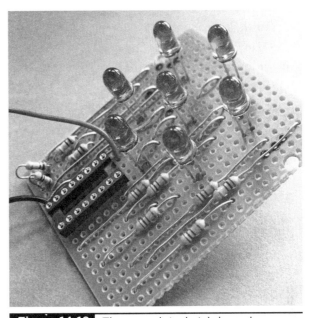

Figure 14-10 The completed stripboard

After you have built the board, test its operation by applying a 3V supply to the board (without IC1 fitted) and checking the voltage levels at pins 8 and 16. Activate each of the seven IR LEDs individually by applying a positive voltage to pins 9 to 15 in turn, but remember that you will need to view the LEDs through a digital camera to make sure that they illuminate.

Once you are happy with the board layout, remove the 3V supply and fit the 4511B device into the DIL socket. Then reapply power to the board and rotate SW2 to ensure that you see the coded shapes outlined in Figure 14-2 whenever the corresponding number 0 to 9 is selected on SW2. Again, you will need to use a digital camera to see these images.

Putting It All Together

Now you can fit the stripboard into the enclosure. I used only two of the four fixing holes, which provided sufficient support to hold the board in place. Next, insert push button SW1 into the side of the enclosure lid. Then, use double-sided adhesive to affix the battery holder to the inside of the lid. Finally, cut the positive battery lead and solder the two ends to SW1, and use a cable tie and a fixing base to secure the battery leads to the lid. You should end up with a completed unit that looks like the one shown in Figure 14-11. You can now fit the enclosure lid to the base and screw it all together.

The enclosure that I used has a removable battery compartment, which makes it nice and easy to change the batteries when required. The rear of the finished enclosure with the battery compartment lid removed is shown in Figure 14-12.

Send Me a Message

Now for the fun part. Give the device to one of your friends and ask them to stand across the room from you. Tell them to select a random number on the rotary switch and then to press and hold the transmit button while you view the display using a digital camera. You should quickly be able to recognize the number shape by comparing it to

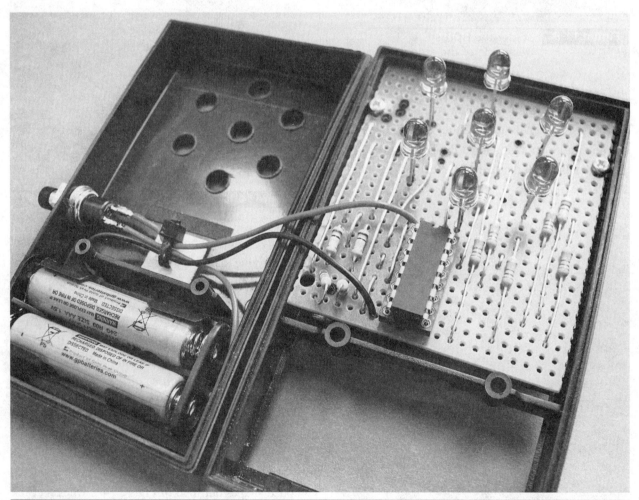

Figure 14-11 The completed device

Chapter 14 ■ A Manually Operated Sequencer: Invisible Secret Code Display

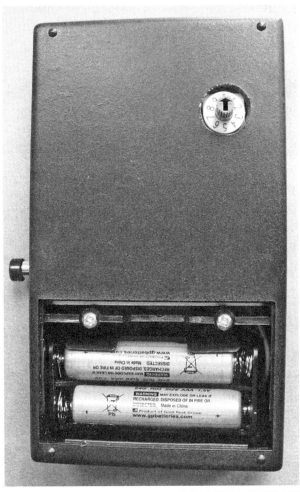

Figure 14-12 The rear view of the completed enclosure

that each letter of the alphabet is represented by a two-digit number:

A = 01

B = 02

C = 03, etc.

You can then send a sequence of secret shapes that, when deciphered, creates a message. So, for example, the word "Hi" would have a numerical code of 0809. The sender would therefore rotate the BCD switch to each number in turn, and then hold the transmit button (SW1) for a couple of seconds so that the person receiving the message can identify the coded shape and decipher the number and write it on a pad. In this example, the person transmitting the letter H would follow this sequence:

1. Rotate the BCD switch to number 0.
2. Press the transmit button for 2 seconds.
3. Release the transmit button.
4. Rotate the BCD switch to number 8.
5. Press the transmit button for 2 seconds.
6. Release the transmit button.

those shown in Figure 14-2 and tell your friend which number they selected on the rotary switch. Hopefully, they will look amazed. In fact, this little trick in itself could form the basis of a magic routine. You may need to use the zoom feature on the camera to see the shape clearly, depending on how far away you are from the other person. This effect also works better in the dark than in normal daylight. Anyone who is looking at the display without a digital camera would not know that the coded shape is even there.

You could use this device to transmit a secret message to your friend by using a simple alphanumeric code. For example, you might decide

It's up to you to decide how easy or difficult to make the secret code, and you might decide to arrange a particular flash sequence to show the person receiving it that the message is about to start. In fact, you could use the device to flash out a Morse code sequence by setting the BCD switch to number 8 so that all seven LEDs are lit, and then you could press the transmit button in a dot-dash sequence. Just remember that if you aren't viewing the display using a digital camera, you can't see what is being transmitted. An example of the number 8 being transmitted by the device is shown in Figure 14-13.

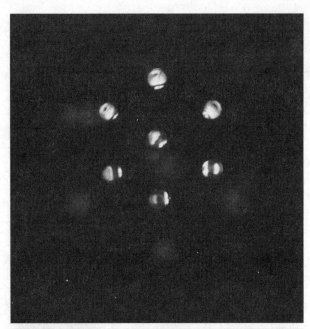

Figure 14-13 The number 8 viewed through a digital camera

Future Modifications

The series resistors used to drive the LEDs are set so that they are visible in the dark. If you want to make the IR light output brighter, so that you can see the images better in daylight, you will need to take into consideration the maximum output current capabilities of IC1. The data sheet for the 4511B device that I used suggests that the output current available for driving LEDs is 25mA (please refer to the relevant manufacturer's data sheet for the one that you use for more information). The resistor values suggested for R5–R11 in the parts list of this project may already be pushing the limits of IC1 when all seven LEDs are illuminated, but don't forget that the display should only be illuminated for a few seconds at a time whenever you push the transmit button. Having tested the circuit over a number of hours, my prototype worked fine and didn't seem to have any detrimental effect on IC1. You may prefer, however, to increase the resistance values of R5–R11 in your project to avoid any potential damage to IC1.

Be aware that if you decide to reduce the resistance values of R5–R11, which will increase the current through each LED and increase their brightness, this could damage IC1, and also remember that SW1 carries the overall current of the complete circuit. If you want to make the IR LEDs a lot brighter and more visible over a longer distance, you will need to re-design the circuit to include some additional LED driver circuitry (similar to the Color-Changing Disco Lights project in Chapter 9), and you may also need to change switch SW1 to one with a higher current capability, or possibly use the "blanking" facility of the 4511B to activate the LEDs.

PART THREE
POV Projects

CHAPTER 15

Basic LED Matrix and POV Concepts: How to Build a Three-Digit Counter

THE PROJECTS IN THE PRECEDING TWO PARTS OF THIS book demonstrate how to create some interesting effects using various types of LEDs either by illuminating them or flashing them in a particular sequence. The projects presented in this third and final part of the book show you how to build circuits that create "persistence of vision" (POV) effects. This chapter first explains how the POV effect works and briefly describes LED multiplexing circuit principles. The chapter's project then demonstrates how a typical matrix circuit can be made to produce the POV effect by using three seven-segment displays to create a three-digit counter, shown in Figure 15-1. You could house the board in an enclosure to make a neat handheld people counter, the type that you see festival or concert staff using to count the number of people entering a venue. In fact, you could use it for any counting application where you need to count up to 999.

Figure 15-1 The three-digit counter board

Persistence of Vision (POV)

Persistence of vision (POV) is an effect that we are all familiar with, probably without realizing it. It describes the optical illusion whereby you continue to see an image for a split second after you view it, often referred to as an *afterimage* or *ghost image*. To experience this effect, stare closely at a black shape or image printed on a piece of white paper for around 30 seconds, and then immediately look away at a light-colored wall. This should give the illusion of the same shape or image appearing in white when you look at the wall. It is commonly thought that this effect also occurs when you watch a film at the cinema; the film reel contains a sequence of slightly changing static images projected onto the screen and presented to your eyes at approximately 25 frames per second. Because of the POV effect, your eyes retain the previous image before the next one appears, the effect of which is that you see a continuous moving image without seeing the gaps in between each of the images. It is debated, however, that it is not actually POV that causes us to see moving images in this way, and if you are interested you can find a lot more information about this subject on the Internet. The projects that you will read about in this section of the book are classified as POV projects, and they use some very fast flashing LED techniques to create some exciting visual illusions.

LED Multiplexing Circuit Principles

You can reproduce the POV effect using LEDs to create some really interesting projects. The benefit of using LEDs for this type of project is that they can be switched on and off very quickly, which is perfect for creating the POV effect. You can also reduce the overall current consumption of the POV circuits because you can build your circuits in a matrix format, which allows you to multiplex the LED outputs. In the forthcoming projects, you will see various circuit layouts that use a similar multiplexing principle to create the desired POV effect. The upcoming "How the Circuit Works" section provides more details.

Project 14
Building the Three-Digit Counter

If you have already built the Mini-Digital Display Scoreboard project in Chapter 7, you will realize that this project could be built by using three 4026B ICs, and 21 series resistors to limit the current the three seven-segment displays. The project specifications show that this project uses a microcontroller instead, which allows you to see POV in action, while also reducing the component count. Before building this project it is recommended that you read Chapter 12, which outlines the key features of the PIC16F628-04/P microcontroller and explains how to program this device.

PROJECT SPECIFICATIONS

- Three seven-segment LEDs create a three-digit counter display.
- Microcontroller-generated multiplexed circuitry creates a persistence of vision effect.
- The three-digit counter has a single push button count input.
- The switch debounce and reset facility is enabled via software.
- The supply voltage is 3 volts.

How the Circuit Works

Figure 15-2 shows the circuit diagram for the three-digit counter project.

The purpose of this circuit is to produce a three-digit display that is capable of keeping count from 000 to 999, increasing by one each time the count button is pressed. Before I describe how the circuit works, consider the current consumption of such a circuit if all the LEDs in all three displays are illuminated, including the decimal point (DP). Assuming that each segment is set to draw 6mA, multiply that by 8 (seven segments plus the DP), and then multiply 48mA by 3 (three digits); the result indicates that the display alone will draw 144mA, which is a high current draw for a battery-powered circuit, meaning the batteries would not last very long. Take a look at the circuit diagram in Figure 15-2 and you will see that the PIC

Chapter 15 ■ Basic LED Matrix and POV Concepts

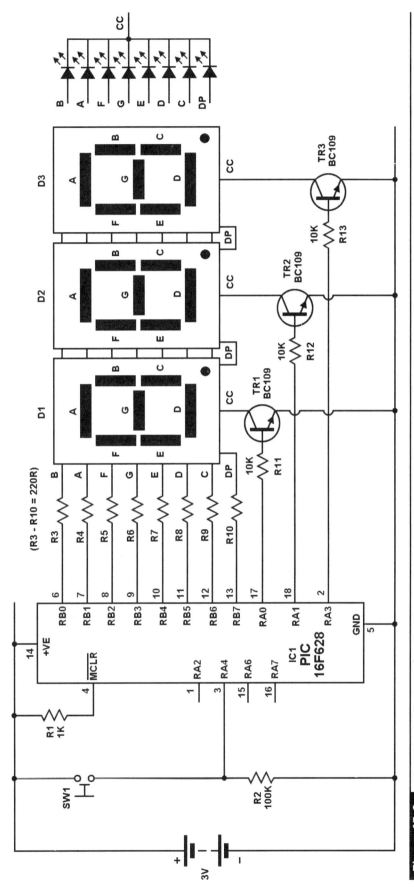

Figure 15-2 The circuit diagram for the three-digit counter

Microcontroller (IC1) has all eight outputs of Port B connected to the eight anodes of the first seven-segment display (D1) via series resistors R3 to R10, and those connections are then linked to the eight anodes of D2 and also to those of D3. Now look at the cathode connections from each of the three displays (D1 to D3); these are not hard-wired to the ground rail of the circuit as expected, but instead are fed through transistors TR1 to TR3, which are individually controlled by three outputs on Port A of IC1 (RA0 to RA3). You will soon see how this type of circuit configuration is able to vastly reduce the current consumption of the circuit.

Now I'll explain how the circuit works. The software that is programmed into IC1 makes sure that only one seven-segment display is on at any one time. Each of the transistors is made to switch on and off in sequence, starting at TR1, which activates D1 for a split second and then is switched off; TR2 follows, activating D2 for a split second before being switched off; and finally TR3 is switched on to activate D3 for a split second and is switched off. This sequence continues in a never-ending loop. As this switching sequence occurs, the eight-digit binary code for the relevant display is presented to Port B in turn. So, for example, when TR1 is activated, Port B outputs the binary code for LED D1, then when TR2 is activated, Port B outputs the code for LED D2, and, finally, when TR3 is activated, the code for LED D3 is presented. Because all of the anode connections of the three displays are linked together, only one display is activated at a time, depending on which transistor is activated to allow the current to flow through the LEDs to the ground rail. This switching sequence is performed at a very fast rate, which creates a POV effect and makes it look as though all three displays are switched on at the same time.

As mentioned earlier, the other benefit of this type of multiplexed circuit layout is that the overall current consumption is reduced because only one seven-segment display is illuminated at a time. For example, if all LED segments are lit to show the number 8.8.8. on the counter, the current consumption of the three-digit display would be reduced to around 48mA, which is one-third of the original display consumption of 144mA. In fact, the overall current consumption may be slightly less than this because the displays are being switched on and off at a very fast rate.

SW1 and R2 form an input circuit, which is connected to Port A,4. This circuit is normally held low via R2, so the software recognizes when the button is pressed because pin 3 (RA4) receives a high signal.

NOTE You will see that the circuit is powered via two AA batteries, which provide a 3V supply to power the PIC16F628 microcontroller (IC1). According to the datasheet for the PIC16F628-04/P device, the lowest voltage that it will accept is 3 volts; however, experimentation has proven that the device will often work at voltages slightly lower than this. There is also a lower voltage version of this microcontroller called the PIC16LF628; you might want to investigate the specifications of this device for your own project designs.

The Display Codes

To be able to generate digits 0 to 9 on the seven-segment displays I had to create a binary look-up table in the software that I wrote for this project; this was created by using the values shown in Table 15-1. Columns B0 to B7 represent the output pins of Port B, and the letters in brackets represent the segment letter of the seven-segment display that each is connected to. This coding concept is similar to how the 4026B CMOS device generates its output to drive a single seven-segment display, as described in Chapter 7. A value of 1 in the table denotes that the segment is lit, and a value of 0

TABLE 15-1 The Binary-Coded Sequence to Generate Numbers 0 to 9

Number to be Displayed	B7 (DP)	B6 (C)	B5 (D)	B4 (E)	B3 (G)	B2 (F)	B1 (A)	B0 (B)
0	0	1	1	1	0	1	1	1
1	0	1	0	0	0	0	0	1
2	0	0	1	1	1	0	1	1
3	0	1	1	0	1	0	1	1
4	0	1	0	0	1	1	0	1
5	0	1	1	0	1	1	1	0
6	0	1	1	1	1	1	0	0
7	0	1	0	0	0	0	1	1
8	0	1	1	1	1	1	1	1
9	0	1	1	0	1	1	1	1
L	1	0	1	1	0	1	0	0
E	1	0	1	1	1	1	1	0
d	1	1	1	1	1	0	0	1

means that the segment is off. The letters *L*, *E*, and *d* are included in the look-up table as part of the start-up sequence of the counter.

The benefit of using a microcontroller to generate numerical numbers on a display is that you can customize the coding, which means that you can also create a reduced alphabet character set.

Parts List

NOTE The Supplier and Part Number column of the following table lists specific parts that I used in this project. Refer to the appendix for additional details about acquiring your parts.

The parts you'll need for the three-digit counter project are shown in the following table.

PARTS LIST			
Code	Quantity	Description	Supplier and Part Number
IC1	1	PIC16F628-04/P Microcontroller	RS Components 379-2869 (Mfr: MicrochipTechnology Inc. PIC16F628-04/P)
R1	1	1KΩ 0.5W ±5% tolerance carbon film resistor	—
R2	1	100KΩ 0.5W ±5% tolerance carbon film resistor	—
R3–R10*	8	220Ω 0.5W ±5% tolerance carbon film resistor	—
R11–R13	3	10KΩ 0.5W ±5% tolerance carbon film resistor	—

(continued)

Code	Quantity	Description	Supplier and Part Number
		PARTS LIST (continued)	
D1–D3	3	Seven-segment red common cathode display (HDSP-5503) V_F (typical) = 2.1V, I_F (typical) = 20mA	RS Components 587-951 (Mfr: Avago Tech. HDSP-5503)
SW1	1	0.24" × 0.24" (6mm × 6mm) tactile momentary push-to-make switch 0.67" high (17mm), 50mA rated	RS Components 479-1463 (pack of 20)
TR1–TR3	3	BC109 NPN transistor (see text)	—
Hardware	1	Stripboard, 0.1" (2.54mm) hole pitch, 37 holes wide by 24 tracks high	—
Hardware	1	18-pin DIL socket	—
Hardware	1	AA battery holder (two AA batteries)	RS Components 512-3580
Hardware	2	AA batteries (1.5V)	—
Hardware	1	PP3 battery clip	RS Components 489-021 (pack of 5)

* Note: If you use seven-segment displays that have different V_F and I_F values to those that are in the parts list, then you may need to alter the resistance and wattage values of these LED series resistors. Please refer to Chapter 2, which explains how to do this, and use a value of 4mA to 6mA for I_F in your calculations for this project. Also read the note at the end of this chapter.

Stripboard Layout

The stripboard layout for the three-digit counter project is shown in Figure 15-3.

Before soldering the components in place, make sure that you make the 30 track cuts that are shown on the stripboard layout as white rectangular blocks, including those that sit underneath the seven-segment displays and IC1.

How to Build and Test the Board

NOTE Please refer to Chapter 1 for soldering tips and techniques and for generic stripboard building guidelines.

Build the stripboard layout by closely following the diagram shown in Figure 15-3. I chose to solder the three displays directly to the board, but you may decide to solder SIL sockets to the board so that you can remove and replace each segment without having to desolder it from the board in the future. Also, notice that one wire link traverses the top of IC1 (link L1), and another wire link is actually soldered to the track side of the stripboard (link L2). You need to ensure that L2 is an insulated piece of wire so that it doesn't short out the tracks, which could cause unstable operation or damage the components.

Once you have built the stripboard, it should look like the board shown in Figures 15-4 and 15-5. The holes shown on the stripboard are not necessarily required; it all depends on how you intend to mount the board in your chosen enclosure.

Before you fit IC1, you can test the operation of the three displays in a similar manner to the tests described in previous chapters, but keep in mind that a display will not illuminate unless its associated transistor is also activated. So, for example, to test each of the individual segments of D1, you will need to link pin 14 (+) of the DIL

Chapter 15 ■ Basic LED Matrix and POV Concepts

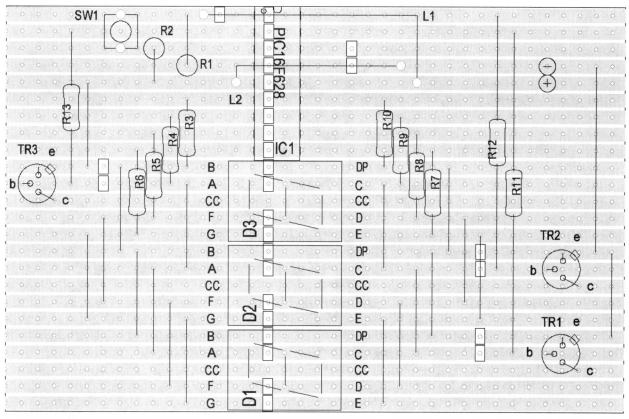

Figure 15-3 The stripboard layout for the three-digit counter

Figure 15-4 The completed stripboard layout for the three-digit counter

Figure 15-5 The track side of the board, showing wire link L2

socket to pin 17 to activate TR1 first. You can then link pin 14 to each segment output pin in turn (DIL socket pins 6 to 13) to prove the operation of each of the eight LED segments in D1. You can repeat the process by activating each transistor in turn, which requires a positive voltage on DIL sockets 17, 18, and 2 in turn. Once you are happy with the board layout, you can move on to programming IC1.

The PIC Microcontroller Program

Now you need to program IC1 and install it on the stripboard. You can download the assembly program and Hex file from the McGraw-Hill website, at www.mhprofessional.com/computingdownload. You can then program IC1 with the Hex file called **LED 3-Digit Counter.hex**, as explained in Chapter 12.

The Assembly Program

The assembly program for this project is called **LED 3-Digit Counter.asm** and includes comprehensive notes to explain how it works. The routine uses look-up tables for the characters that appear on the seven-segment displays. It also has a switch debounce feature to stop spurious triggering when SW1 is pressed, as well as a routine that allows the counter to be reset by holding SW1 down for a few seconds.

The following is an extract of the software showing the seven-segment display character set look-up tables:

```
;7-Segment Display Character set

ALPHA:      addwf PCL,F         ;L.E.d.
                                ;start-up word
                                ;character
                                ;table
            retlw B'10110100'   ;L
            retlw B'10111110'   ;E
            retlw B'11111001'   ;d


CODE:       addwf PCL,F         ;7-segment
                                ;display
                                ;character
                                ;table
            retlw B'01110111'   ;0
            retlw B'01000001'   ;1
            retlw B'00111011'   ;2
            retlw B'01101011'   ;3
            retlw B'01001101'   ;4
            retlw B'01101110'   ;5
            retlw B'01111100'   ;6
            retlw B'01000011'   ;7
            retlw B'01111111'   ;8
            retlw B'01101111'   ;9
```

The Hex File

The Hex file that you need to transfer into IC1 is called **LED 3-Digit Counter.hex**, which is shown next. Program IC1 using this file and insert the microcontroller into the DIL socket on the stripboard that you have built.

```
:020000001528C1
:080008001528152882078B43405
:10001000BE34F9348207773441343B346B344D3489
:100020006E347C3443347F346F3407309F00831642
:1000300010308500003086008130810083120530 49
:10004000A200A00185018601A401A501A601A301CA
:10005000A7010514200806208600 7F208501A00A3C
:10006000851420080620860 07F208501A00A8515BA
:10007000200806208600 7F208501A001051A41285E
:100080002928231405142608 0A2086007F208501CC
:10009000851425080A2086007F20850185152408FF
:1000A0000A2086007F208501051A5D20051EA30118
:1000B0002708C83A031979204228A70A23180800FC
:1000C000A40A24080A3A031968282314A70108007F
:1000D000A401A50A25080A3A031970282314080068
:1000E000A401A501A60A26080A3A031979282314AF
:1000F0000800A501A401A6012314A70108002208F5
:10010000A1008B010B1D82280B11A10B8228080076
:02400E00303F41
:00000001FF
```

POV in Action

After you program IC1 with the hex code and fit it into its DIL socket, fit two AA batteries into the battery holder and connect it to the board to power it up. You should first see the start-up sequence, which is three letters that spell out the word "L.E.d." (see Figure 15-6).

Figure 15-6 The start-up sequence

A lowercase d is used because an uppercase D looks like a number 0 on a seven-segment display. The display is being multiplexed, which means that although only one display is active at any one time, the speed of operation is sufficiently fast to fool the eye into seeing all three displays illuminated simultaneously. If you use a digital camera with a fast shutter speed to view the display, you will see that the display flickers as each segment is switched on and off at a very fast rate.

Now press SW1 briefly and the display should show 000, which means that the software has entered the counter routine, as shown in Figure 15-7.

Figure 15-7 The counter is now operational.

Each time you press SW1, the counter should increment by one until you reach 999, after which the display reverts to 000 again. You can also reset the counter to 000 at any time by holding down SW1 for a couple of seconds.

> **NOTE** The microcontroller circuit layout for this project doesn't contain a decoupling capacitor across the + and − battery supply to smooth the supply voltage and to help avoid potential spurious triggering of the circuit. I didn't include a decoupling capacitor because the circuit worked well without one. If you experience a problem with your project, try fitting a 100nF or 0.1µF (minimum 10V rated) capacitor across the positive and ground rails of the circuit to see if that helps to alleviate the issue.

Possible Enhancements

The beauty of creating generic circuit layouts such as the one in this project is that you can modify the software in the microcontroller to produce an outcome that is totally different from that of the original design. If you are familiar with programming the PIC Microcontroller, you could modify the software in this project so that the counter counts down instead of up. Or you could even modify the software to create a basic slot machine that produces letters, numbers, or shapes on each of the seven-segment displays in turn at the press of the button. The possibilities are almost endless.

> **NOTE** The BC109 NPN transistors used in this project have a maximum collector/emitter current capability of 100mA, which may be exceeded if you make modifications to the circuit. You may need to use transistors with a higher current rating if you alter the series LED resistors or increase the supply voltage to the circuit.

CHAPTER 16

A Multicolor POV LED Circuit: Backpack Illuminator

THE PROJECT IN THIS CHAPTER SHOWS YOU HOW TO make a cool, eye-catching, multicolor flashing display that can also show basic animation sequences. The circuit design incorporates additional driver circuitry to control a matrix of LEDs, and produces fast flashing colorful images to create a POV effect (POV concepts are discussed in more detail in Chapter 15). This chapter's project shows you how you can integrate the display electronics into the fabric of a backpack to help make you more visible outside at night. You could use the same concept to embed a funky electronic badge into your clothing, or as the basis for a project that requires a small, animated display. The illuminator that will be integrated into a backpack in this project is shown in Figure 16-1.

Project 15
Backpack Illuminator

If you have read Chapter 6 you will have seen tricolor LEDs already, and you will realize that they are able to produce three different colors depending on how they are switched. This project puts these multicolor LED devices to good use to create an exciting visual POV effect.

> **PROJECT SPECIFICATIONS**
>
> - The display comprises 16 tricolor LEDs configured in a four-by-four matrix.
> - Each LED can be controlled independently to display one of three colors: red, green, or amber.
> - Colorful flashing images and basic animations can be displayed.
> - The supply voltage is 4.5 volts.

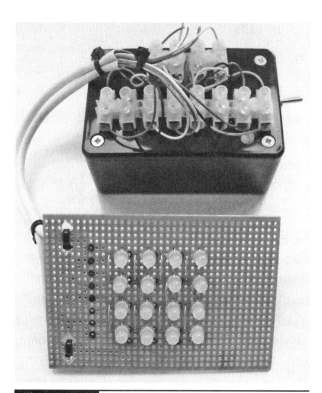

Figure 16-1 Backpack illuminator

How the Circuit Works

The circuit diagram for the backpack illuminator is shown in Figure 16-2. The circuit looks quite complicated, but many of the LED display interconnections are taken care of by the stripboard layout.

The heart of the circuit is IC1, a PIC16F628 microcontroller that is powered by a 4.5V battery. The circuit can be switched on or off via toggle switch SW1. Both Ports A and B of IC1 are configured as outputs, and Port B is connected directly to the 16 tricolor LEDs D1 to D16 via series resistors R2 to R9. The operation of tricolor LEDs, discussed briefly in Chapter 6, is simple: each tricolor LED contains two LED colors, red and green, which allows you to generate up to three colors from each LED, red, green, and amber (amber is created when the red and green elements are illuminated at the same time).

The four-by-four LED matrix in this circuit can be broken down into four vertical columns of LEDs, which are identified as 1C to 4C on the circuit diagram. Consider column 1C first; you can see that the eight outputs of Port B are connected to the anodes (+) of four LEDs (D1, D5, D9, and D13), which means that the 8 bits of data that are generated from Port B can individually control each color lead for all four LEDs. The cathode (−) connections of these four LEDs are connected together to create column 1C. This connection structure is repeated for all four columns of LEDs (1C to 4C). The four LED cathode columns are not connected directly to ground as you might expect; instead, they are actually switched on and off one at a time in a sequence generated by the outputs of Port A so that only one column of four LEDs is illuminated at a time. The purpose of this matrix structure is to enable you to individually control each of the 16 LEDs using 12 outputs from IC1.

Because the maximum drive capability of each port output of IC1 is 25mA, if you were to rely on each of the four outputs of Port A to sink a 20mA current of four tricolor LEDs at a time, the port would be trying to sink $8 \times 20mA = 160mA$. This means that you need to provide some current-boosting circuitry so that you can sink the current from each LED column. I did originally consider using four individual transistors to do this, but I then decided to use a neater method and utilize the ULN2003 device (IC2), which contains seven transistor drive circuits instead; only four of the seven drivers are used in this design. Port A outputs A0 to A3 feed the base input connections of IC2 (1B to 4B), and the collector outputs are fed to the four LED cathode banks (1C to 4C).

The matrix display is controlled by software that needs to be programmed into IC1, and this operates in the following way:

1. Present 8 bits of data to Port B.
2. Activate RA3 to display the 8 bits on column 1C LEDs.
3. Deactivate RA3.
4. Present 8 new bits of data to Port B.
5. Activate RA2 to display the new 8 bits on column 2C LEDs.

This process continues until all 16 LEDs have been illuminated and repeats in a never-ending loop. Only one column of four LEDs is illuminated at any one time, and this happens at a fast enough speed to fool the eye into seeing all 16 LEDs illuminated at the same time. If you understand how each bit of data operates the individual color elements of the LEDs, you can create colorful moving and flashing images, as explained in more detail later in the chapter.

Chapter 16 ■ A Multicolor POV LED Circuit: Backpack Illuminator 175

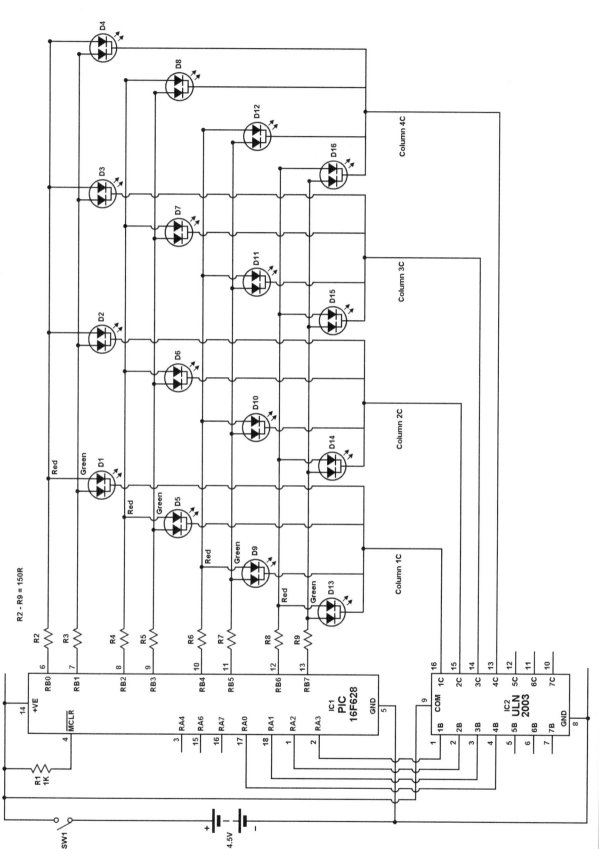

Figure 16-2 Circuit diagram for the backpack illuminator

Parts List

> **NOTE** The Supplier and Part Number column of the following table lists specific parts that I used in this project.

Refer to the appendix for additional details about acquiring your parts.

Here are the parts you'll need for the backpack illuminator project.

PARTS LIST

Code	Quantity	Description	Supplier and Part Number
IC1	1	PIC16F628-04/P Microcontroller	RS Components 379-2869 (Mfr: MicrochipTechnology Inc. PIC16F628-04/P)
IC2	1	ULN2003AN Darlington transistor array	RS Components 436-8451
R1	1	1KΩ 0.5W ±5% tolerance carbon film resistor	—
R2–R9*	8	150Ω 0.5W ±5% tolerance carbon film resistor	—
D1–D16	16	5mm tricolor LED Red LED V_F (typical) = 2.0V, I_F (max)= 30mA Green LED V_F (typical) = 2.2V, I_F (max)= 30mA	ESR Electronic Components 715-250
SW1	1	Single-pole panel mount toggle switch, 2A rated	RS Components 710-9674
Hardware	2	Stripboard, 0.1" (2.54mm) hole pitch, 37 holes wide by 24 tracks high	—
Hardware	1	18-pin DIL socket	—
Hardware	1	16-pin DIL socket	—
Hardware	1	AA battery holder (three AA batteries)	Maplin YR61R
Hardware	1	PP3 battery clip and lead	RS Components 489-021 (pack of 5)
Hardware	3	AA battery (1.5V)	—
Hardware	1	Enclosure, approx. 3.3" (L) × 2.2" (W) × 1.6" (H) (85mm × 56mm × 40mm)	RS Components 281-6879
Hardware	—	12-way terminal block	Electrical hardware store
Hardware	—	Cable ties, cable tie base, M3 nylon screws and nuts, double-sided adhesive strips, 12-core flexible cable (with various wire colors, each core rated at least 200mA)	—
Material	—	Backpack, weatherproof fabric (optional; see text)	—

* Note: If you use LEDs that have different V_F and I_F values to those that are in the parts list, then you may need to alter the resistance and wattage values of these LED series resistors. Please refer to Chapter 2, which explains how to do this, and also refer to Chapter 12, which outlines the current capabilities of the PIC16F628-04/P microcontroller. The value of I_F in your calculations for this project should be lower than 20mA.

Stripboard Layout

This project has two stripboard layouts: the driver board contains the bulk of the electronic driver circuitry (see Figure 16-3), and the display board contains the 16 LEDs to create the matrix display (shown later in the chapter in Figure 16-7). The two boards are connected together using just 12 interconnecting cables. Think about that for a second—you can control 16 individual tricolor LEDs, each containing two colors, using just 12 control wires. Bearing in mind that each LED can have one of four possible states (off, red, green, or amber), this means that if you had enough memory in the microcontroller, you could in theory generate 4,294,967,296 different display combinations.* This is the beauty of using matrix displays.

The driver board is 24 holes wide by 19 tracks high. I cut down a piece of stripboard that was originally 37 holes wide by 24 tracks high. Make sure that you make the 25 track cuts shown in Figure 16-3, some of which sit underneath the IC sockets and resistors (track cuts are shown as white rectangular blocks). You also need to drill two holes so that you can mount the board onto the inside lid of the enclosure.

The display board that I used is 37 holes wide by 24 tracks high, but you could make this smaller if you prefer a smaller display. Note in Figure 16-7 that you need to make 15 track cuts and then drill four holes to secure the interconnecting cables to the board.

How to Build the Driver Board

NOTE Please refer to Chapter 1 for soldering tips and techniques and for generic stripboard building guidelines.

Build the stripboard layout by following the diagram shown in Figure 16-3. Once you have

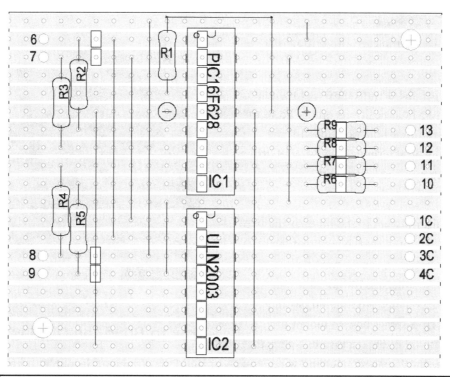

Figure 16-3 The driver board layout for the backpack illuminator

---

* The number of display combinations is calculated as follows: Each LED column = 8 bits of binary data = 256 possible combinations per column. There are four LED columns, so this gives you a total of 256 × 256 × 256 × 256 = 4,294,967,296 different display combinations.

built the board, you can solder 12 color-coded interconnecting cables to the points marked 6 through 13 and 1C through 4C in Figure 16-3. These cables need to be capable of carrying at least 200mA each, and should be long enough to feed through a hole in the enclosure lid and reach two terminal blocks on the outside of the lid. The completed driver board should look like the one shown in Figure 16-4.

Figure 16-5 Use nylon screws and nuts to mount the board onto the lid.

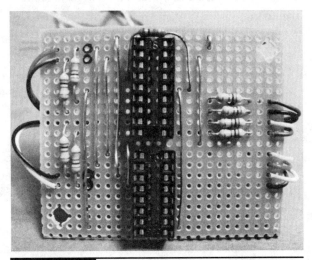

Figure 16-4 The completed driver board

Cut the cables to size, remove their insulation, and screw them into the two terminal blocks. Use the eight-way block for cable connections 6–13 and the four-way block for cable connections 1C–4C. Figure 16-6 shows the lid layout from the outside with the 12 cables cut to size and screwed into the terminal blocks.

Figure 16-6 The completed enclosure lid

You can now use the driver board as a template to mark out the two hole positions on the lid of the enclosure, mark out four hole positions for the eight-way and four-way terminal blocks, and drill a hole near the center of the lid through which you can feed the 12 cables.

Next, affix the stripboard to the lid by using two M3 nylon screws and nuts. Make sure that the stripboard is positioned in such a way that the board does not touch the side of the enclosure when you fit the lid, which could stop the lid closing properly. Figure 16-5 shows how to mount the stripboard onto the lid of the enclosure. Notice that you need to place the 12 interconnecting wires under the stripboard layout and feed them through the central hole in the lid (as shown in Figure 16-6).

How to Build the Display Board

Build the stripboard layout by following the diagram shown in Figure 16-7. Solder the solid wire links in place first and then solder the 16 tricolor LEDs to

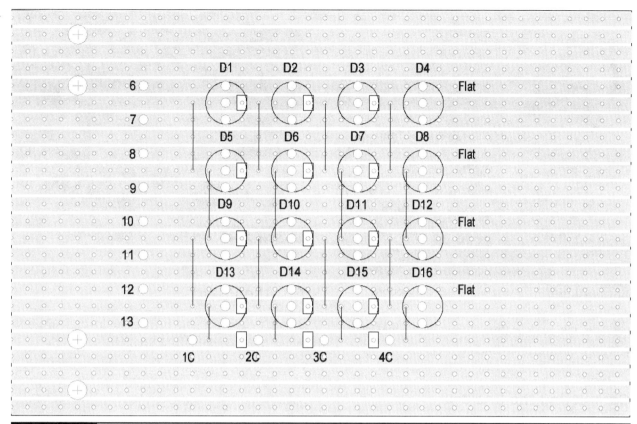

Figure 16-7 The display board layout for the backpack illuminator

the board, taking note of the orientation of the flat side of the LEDs, which is shown on the diagram. After you solder the 16 LEDs in place, solder 12 color-coded interconnecting cables to the copper side of the board, as shown in Figure 16-8. Match the color coding that you used for the driver board for the points marked 6 through 13 and 1C through 4C.

The length of the interconnecting cables depends on how far from the display board you intend to mount the driver board. In my project, the cables are only about eight inches long because I mounted the controller very close to the display board. Once you have soldered the 12 wires in place, secure them to the board using two cable ties through the four holes that you drilled into the stripboard. This helps to prevent the cables from being pulled and damaged. The completed LED display board should look like the one shown in Figure 16-9.

Now cut the 12 core cables to size and connect them to their corresponding colors that are fitted in the eight-way and four-way terminal blocks on the enclosure lid. Take your time to make sure that these interconnecting cables are in the correct

Figure 16-8 Solder the 12 cables to the copper side of the board.

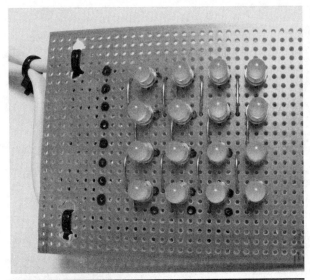

Figure 16-9 The completed display board

Figure 16-10 The layout of the enclosure

position; taking your time at this stage could save you a lot of time fault-finding later.

Completing the Enclosure

Fit the PP3 battery clip to the AAA battery holder and use double-sided adhesive to affix the battery holder to the base of the enclosure. You may need to sand one edge of the PP3 clip slightly to allow it to fit snugly against the fixing posts. Once it is in place, mark out a suitable location for the power switch SW1, drill a hole in a narrow side of the enclosure, and secure the switch in place. Cut the positive lead of the battery clip and solder the two ends to the switch, and then secure the battery cables to the side of the enclosure using a cable tie base. You can then solder the red and black cables to the + and − positions on the driver board, as shown in Figure 16-3. The inside of the enclosure should now look as shown in Figure 16-10. Do not fit the batteries at this stage; the batteries are only shown in Figure 16-10 for clarity.

The completed backpack enclosure should look similar to the one shown in Figure 16-11.

Test the Boards

Before you fit IC1 and IC2 to the driver board, you might want to apply the 4.5-volt battery to the circuit and test that the voltage points on both DIL sockets are as expected by following the circuit diagram in Figure 16-2, and using your multimeter to check out various voltage points. For example, IC1's DIL socket pins 4 and 14 should be +4.5 volts, when you connect the negative connection of your multimeter to pin 5 of IC1's DIL socket. IC2's DIL socket pin 9 should be +4.5 volts when you connect the negative connection of your multimeter to pin 8 of the IC2's DIL socket.

You can also check the operation of each LED color, but you will need to fit IC2 to the driver board first, and make sure that a positive voltage supply is applied to each column input and each LED, thus simulating the operation of Port A and Port B of IC1 when it is operational. So, for example, to check that D1 illuminates red, you need to apply a positive voltage to pin 6 of the DIL socket for IC1, and at the same time apply a positive voltage to pin 2 of the DIL socket for IC1. Work through the circuit diagram carefully and make sure that all LED colors operate as expected. As mentioned earlier, if you can ensure that the circuit operates as expected before installing IC1,

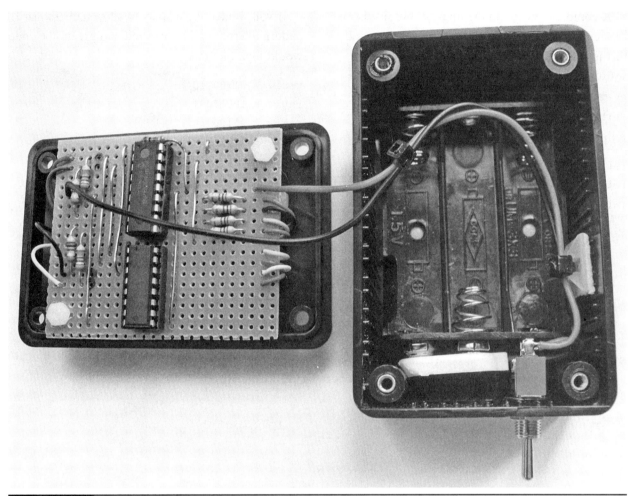

Figure 16-11 The completed backpack illuminator enclosure

you will save yourself a lot of time later. Once you have finished testing the boards make sure that the batteries are disconnected from the circuit before programming and fitting IC1.

The PIC Microcontroller Program

You now need to program IC1 and install it on the stripboard. You can download the assembly program and Hex file from the McGraw-Hill website, at www.mhprofessional.com/computingdownload. You can then program IC1 with the Hex file called **LED Backpack Illuminator.hex**, as explained in Chapter 12.

The Assembly Program

The assembly program listing that you can download for this project is called **LED Backpack Illuminator.asm** and contains plenty of notes that help to explain how it works. The animation sequences are contained within the table called CODE. Colors can be generated on each LED by altering each 2-bit sequence of the 8-bit data that is output on Port B as follows:

- 00: LED off
- 01: Red
- 10: Green
- 11: Amber

A complete image on the four-by-four display is created by generating 32 (4 × 8) bits of data, and this is contained in the look-up table in the software. So, for example, if you wanted the four-by-four display to illuminate amber on all LEDs, the binary code required to do this would be as follows:

```
retlw B'11111111'
retlw B'11111111'
retlw B'11111111'
retlw B'11111111'
```

If you wanted the display to show a chessboard pattern alternating between red and green, the code would look like this:

```
retlw B'01100110'
retlw B'10011001'
retlw B'01100110'
retlw B'10011001'
```

The majority of the program is taken up by the animation look-up table. The complete program listing is too large to show here, so here is the heart of the program, which generates the images:

```
        movlw 30           ;reducing this value makes the animation run quicker
        movwf REPEAT       ;it is the number of times each 4-byte frame is repeated

START:  movlw 58           ;this is the total number of frames in the animation
        movwf FRAME        ;limit the number to 58

        clrf DISPL
        clrf MAP
        clrf COUNT

ST:     movlw B'00001000'
        movwf COLUMN       ;starts the display sequence at column 1C

        movf MAP,W         ;
        movwf DISPL        ;display mapping starts at 0

ST1:    movf COLUMN,W      ;move column to w
        movwf PORTA        ;activate port A column

        movf DISPL,W       ;move display variable into w
        call CODE          ;calls display byte from lookup table
        movwf PORTB        ;moves the lookup value to port B
        call PAUSE         ;pause to stabilize the display
        clrf PORTA         ;clears port A to de-activate columns 1c to 4c
        incf COUNT,W       ;increments count variable
        xorlw 4            ;does count = 4?
        btfsc STATUS,Z
        goto ST2           ;yes, a frame is complete
        incf COUNT,F       ;no, increment count
        incf DISPL,F       ;no, increment display variable
        rrf COLUMN,F       ;rotate the column position to the right
```

```
            goto ST1            ;go back to the start again

ST2:    movf MAP,W              ;move map to w
        movwf DISPL             ;move w to display variable
        clrf COUNT              ;clear count
        decfsz REPEAT,F         ;decrease repeat, has it reached zero
        goto ST                 ;no, keep showing the same frame

NEXT:   movlw 30                ;yes, prepare to show the next frame
        movwf REPEAT            ;sets the repeat value at the default value again
        movlw 4
        addwf MAP,F             ;adds 4 to the map to shift to the next frame
        decfsz FRAME,F          ;have all animation frames been shown?
        goto ST                 ;no, go and show the next 4-byte frame is shown

        goto START              ;yes, show the animation from the start
```

The Hex File

The file for the hex code that you need to program into IC1 is called **LED Backpack Illuminator.hex**. The complete hex code is shown here:

```
:02000000EF28E7
:08000800EF28EF288207AA345B
:10001000AA34AA34AA34AA34AA34AA34AA34EA34B0
:10002000AA34AA34AA34FA34EA34AA34AA34FE34FC
:10003000FA34EA34AA34FF34FE34FA34EA34FF34B2
:10004000FF34FE34FA34FF34FF34FF34FE34FF341F
:10005000FF34FF34FF34FF34FF34FF34FF34FF3408
:10006000FF34FF34FD34FF34FF34FD34F534FF3406
:10007000FD34F534D534FD34F534D5345534F53408
:10008000D53455345534D53455345534553428
:10009000055345534553455345534553418
:1000A000553455345534553469346934553446
:1000B000D734D734FF34553469346934553478
:1000C000D734D734FF34553469346934553412
:1000D000553455345534663499346634993499934EA
:1000E000663499346634663499346634993474
:1000F000663499346634663499346634993464
:10010000663499346634AA34AA34AA34AA34AA3498
:10011000AA34AA34AA34FF34EB34EB34FF34FF346E
:10012000FF34FF34FF34FF34EB34EB34FF34553409
:10013000055345534553455345534FF34D734D734FF34AA34CA
:10014000AA34AA34AA34FF34EB34EB34FF346934D4
:100150000963496346934EB34BE34BE34EB346934AF
:100160000963496346934EB34BE34BE34EB3469349F
:100170000963496346934EB34BE34BE34EB340034F8
:100180003C343C340034FF34C334C334FF340034D3
:10019000014341434003455344134413455340346B
:1001A000283428340034AA3482348234AA34003407
:1001B0003C343C340034FF34C334C334FF340034A3
:1001C000014341434003455344134413455340343B
:1001D000283428340034AA3482348234AA340730D4
:1001E0009F00831600308500003086008030810003B
:1001F00083120530A200850186011E30A4003A302A
:10020000A700A001A501A6010830A3002508A000B1
:100210002308850020080620860023218501260A60
:1002200043A03191729A60AA00AA30C08292508CD
:10023000A000A601A40B04291E30A4000430A507C9
:10024000A70B0429FF282208A1008B010B1D2629DA
:080250000B11A10B2629080087
:02400E00303F41
:00000001FF
```

Once you have programmed the code successfully into IC1, you can then fit the IC into the DIL socket on the driver board.

Testing Time

After you have installed IC1, you can insert the three batteries into the holder and then switch the unit on. If everything is working correctly, you should see a colorful flashing animation sequence. If this is not the case, then you need to do some fault-finding. If you have verified the operation of the LEDs and IC2 before installing IC1, then try to reprogram the microcontroller to see if that is where the problem lies. The flashing animation sequence is generated by the code in the look-up table.

NOTE If you are familiar with programming microcontrollers, try modifying the sequence to create your own colorful animations.

NOTE The microcontroller circuit layout for this project doesn't contain a decoupling capacitor across the + and – battery supply to smooth the supply voltage and to help avoid potential spurious triggering of the circuit. I didn't include a decoupling capacitor because the circuit worked well without one. If you experience a problem with your project, try fitting a 100nF or 0.1μF (minimum 10V rated) capacitor across the positive and ground rails of the circuit to see if that helps to alleviate the issue.

Incorporating the Display into Fabric

For this project, I decided to incorporate my LED display into a backpack so that it would provide a colorful and eye-catching illuminator to help me to be seen in the dark. You can incorporate it into any fabric you want by using similar methods. One of the reasons that I designed this circuit with a separate display board to hold the LEDs was to make sure that most of the sensitive electronic components, installed in the enclosure, can be mounted in a dry location away from the display. Another reason is that it results in no bulky components, which is ideal for incorporating the display into fabric.

Even though LEDs are encapsulated in plastic, you do not want the stripboard to get wet. Therefore, you need to take some measures to try to protect the board from moisture. First of all, you could buy a special waterproofing spray for electronics and spray it on the display board. This will help to protect the board, but it's not a complete solution. I considered using a clear enclosure to mount the display, but I wanted the display to blend into the fabric of the backpack, so I decided to incorporate it into some black fabric. I found an old black anorak made out of waterproof material and I cut out a piece of fabric that was larger than the display board. I then marked the positions of the 16 LEDs onto some tracing paper and transferred this layout onto the fabric. I then used a special hole-punching tool and a hammer to make sixteen 5mm-diameter holes in the fabric, as shown in Figure 16-12. I used a cutting mat and placed a piece of card underneath the fabric so that the hole puncher created neat cuts in the fabric.

Figure 16-12 Punch 16 holes into the waterproof fabric.

CAUTION Even though hardly any heat is generated by the LED display, you still need to use a nonflammable fabric material to encapsulate the display.

After you have punched the holes into the fabric, you can simply push the 16 LEDs through the fabric so that they protrude. Before you do this however, you need to cut a thicker piece of fabric and sew it onto the copper side of the board, using the holes in the board to help to sew it into place (see Figure 16-13). This is to protect the outer fabric from being pierced by the solder joints.

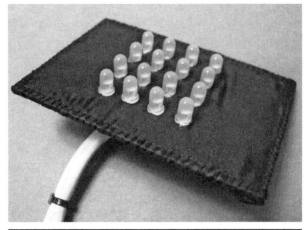

Figure 16-14 Create a fabric pocket for the display.

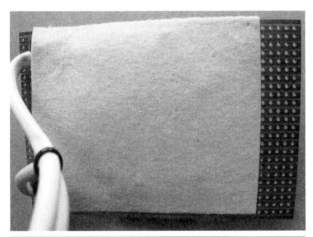

Figure 16-13 Sew a thick piece of fabric to the rear side of the board.

Then sew the waterproof fabric around the stripboard, using the holes in the stripboard again to make a neat stitch, to create a fabric pocket as shown in Figure 16-14.

Now sew the pocket up and cut a hole in the backpack to allow the 12-core display cable to feed through into the backpack. You can either sew the fabric pocket into place on the backpack or use hook and loop strips to hold it in place. You can install the driver enclosure inside a pocket of the backpack and feed the interconnecting display wires into the backpack and connect them to the driver terminal blocks. I would not suggest that you use this method if the backpack is going to be exposed to heavy rain, as water will eventually penetrate into the display board and could eventually damage it.

The illuminator installed on my backpack is shown in Figure 16-15, and the illuminated display is shown in Figure 16-16. The black waterproof

Figure 16-15 The completed backpack

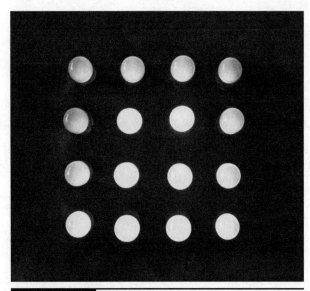

Figure 16-16 The illuminated display

fabric provides a good contrast to the bright colors that are generated by the display and enhances the visual experience.

CAUTION Carrying this type of modified backpack on public transport isn't advisable. Fellow passengers and relevant authorities will probably be suspicious of a bag that contains electronic circuitry, and may not appreciate your love of electronics! I only use my backpack for walking and when riding my bike.

Other Ideas

You could use a similar method to create an electronic fabric badge and sew it onto your clothing, such as on the front of a T-shirt. If you do this, make sure that the fabric pocket has a thick piece of fabric or plastic fitted behind the stripboard to ensure that the solder joints and leads don't pierce the fabric and cut into your skin.

CHAPTER 17

Using a Dot-Matrix Display to Show a Waveform: Digital Oscilloscope Screen

AN OSCILLOSCOPE (OR SCOPE) IS A PIECE OF TEST and measurement equipment that is used by electronics engineers to enable them to visualize and measure voltage signals and waveforms in an electronic circuit, such as square, sine, and triangle waves. The equipment is mainly used for fault-finding and designing prototype electronic circuits. These complex pieces of equipment normally display their waveforms on a small cathode-ray screen (or, more recently, LCD screen) and are able to measure high-frequency signals into the megahertz (MHz) range.

The project presented in this chapter is an experimental circuit that demonstrates how a constantly changing voltage can be used to create a persistence of vision (POV) effect on a dot-matrix display. It also shows how you can emulate the operation of an oscilloscope screen. This circuit does not profess to be an equivalent of a standard oscilloscope screen, because it has a low screen resolution and measures frequencies only up to 100 hertz (Hz), but it could be used as the basis of a simple digital oscilloscope or even a sound-to-light disco effect. The completed digital oscilloscope circuit board and screen is shown in Figure 17-1.

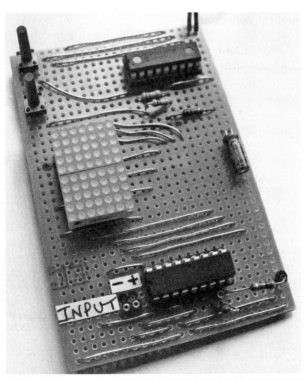

Figure 17-1 The experimental digital oscilloscope screen

Project 16
Digital Oscilloscope Screen

This is the first project in this book that utilizes a dot-matrix LED display as a visual output; in fact, this project uses two displays to create a single display containing 70 LEDs. You will soon see that using these types of displays can vastly reduce the amount of soldering required in your project designs. The project also demonstrates how to create a POV effect by showing a moving voltage signal on the LED display, by using just two integrated circuits and a handful of components. Chapter 15 explains the concepts of POV in more detail.

NOTE If you have not done so already, you need to build the basic single-LED flasher circuit presented in Chapter 4 to complete this project, because you will be using this circuit to test the operation of the digital oscilloscope screen.

PROJECT SPECIFICATIONS

- The dot-matrix display screen comprises 70 LEDs total, made from two 7 × 5 dot-matrix LED displays.
- The display screen can display a moving voltage input signal as a waveform.
- The waveforms that can be displayed are square, sine, and sawtooth.
- The input signal voltage being measured is configurable from 1.25 volts up to 5 volts.
- The time base speed of the display can be increased or decreased using two push buttons.
- The unit can display signals with frequencies up to 100 Hz.
- The supply voltage is 4.5 volts.

How the Circuit Works

I first designed and built a basic digital oscilloscope using a similar principle to the one described here a few decades ago. Its operation is fairly straightforward. Basically, it is a matrix of LEDs configured such that the X axis of the LED matrix is the time base and the Y axis is the voltage. Imagine that there is a square wave signal being fed into the display circuit. The voltage would vary and would look very similar to the waveform shown in Figure 4-1, back in Chapter 4.

Fortunately, there is a cost-effective integrated circuit that can be used to accept a voltage input and that also allows us to display this as an output on ten separate LEDs. This device is the LM3914 bar-graph display driver, which is a linear output version; there is also a logarithmic output version available, the LM3915. This application uses the linear version, while the logarithmic version tends to be used in audio applications.

The LM3914 can also be configured to illuminate the display LEDs in either bar or dot mode. In bar mode, the ten LEDs illuminate in a similar way to a graphic equalizer on your stereo system, whereas in dot mode only a single LED illuminates at any one time. The device is typically used in circuits to display a voltage such as an analog meter, and it is an ideal IC for this project because it is able to measure fast-moving voltages. In this project's circuit, the ten LED outputs form the basis of the first column of LEDs in the Y axis of your display. As the input voltage increases, the LEDs start to illuminate at output 1 for low voltages and rising to output 10 at the highest voltage.

Have a look at the block diagram for this project's display circuit, shown in Figure 17-2, to

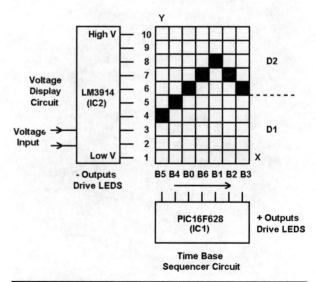

Figure 17-2 Block diagram for the digital oscilloscope screen

see how you can create a display circuit comprising seven columns of ten LEDs.

By activating each column of ten LEDs sequentially, one at a time from left to right, you can start to build up a picture of a moving voltage signal. The voltage of the circuit to be tested is monitored by the input of the LM3914 IC, and its output feeds into the 70-LED display, which is composed of two 7 × 5 dot-matrix LED displays, D1 and D2. The time base sequencer is produced by using a PIC16F628 microcontroller, which activates each column of ten LEDs one at a time in a never-ending sequence. If the sequencer circuit operates at a fast enough speed, the moving voltage that is applied to the LM3914 IC will create a POV effect when displayed on the screen.

In my original design, I used a 555 timer, a 4017 decade counter, and a bank of ten transistors to create the sequencer element of the display. I also used 100 individual LEDs in the display, painstakingly hand-soldering them one at a time to a piece of stripboard to create a 10 × 10 dot-matrix LED display. In this new design, I decided to reduce the component count dramatically by using a PIC Microcontroller to replace all of these components, and I also decided to reduce the amount of soldering required by using two dot-matrix LED displays wired together to create a 10 × 7 dot-matrix display.

This is the first project in the book that uses a dot-matrix display. It is a neat way to avoid having to solder many individual LEDs together to produce an LED display, because the interconnections between each LED are already taken care of inside the display package.

The circuit diagram for the digital oscilloscope screen is shown in Figure 17-3; this should be a lot easier to read now that you understand the display principle.

The circuit is powered using a 4.5V supply in the form of three AAA batteries. This drives IC1, which is a PIC16F628-04/P microcontroller, and IC2, which is the LM3914 bar-graph display driver. The time base (X axis) of the display is generated by IC1. This is created in software by seven Port B outputs which individually switch high, producing a positive output one at a time starting at RB5 and moving through RB4, RB0, RB6, RB1, RB2, and RB3 in turn; this sequence repeats itself in a never-ending loop. These positive outputs activate an LED column of the two common cathode dot-matrix displays (D1 and D2 wired together) one at a time. Ports A2 and A3 of IC1 are configured as inputs in software and monitor switches SW2 and SW3. Pressing these buttons alters the speed of the sequencer: SW3 increases the speed and SW2 decreases the speed. These input pins are held low via resistors R2 and R3 and go high whenever SW2 or SW3 is pressed.

The Y axis of the dot-matrix display is generated by IC2. The voltage input into IC2 (pin 5) is displayed on one of the ten LED outputs by going low. Pin 9 of IC2 is left floating, which configures the device in dot mode, which means that normally only one LED is illuminated at a time. Sometimes this could be two LEDs, because the device has an overlap facility to make sure that there is always an LED illuminated. You will notice that the outputs from IC1 and IC2 do not have any series resistors to limit the current to the LEDs in this circuit. This is because R4 determines the amount of current that flows through the LEDs via IC2, and by looking at the datasheet for the LM3914, you will see that the value of R4 in this circuit can be calculated as follows:

$$R4 = 12.5 / I_{LED}$$

where I_{LED} is the amount of current that we want to flow through the LED.

Because there could be two LEDs illuminated per column, a 2.2KΩ resistor was chosen for R4, which in this circuit limits the current to around

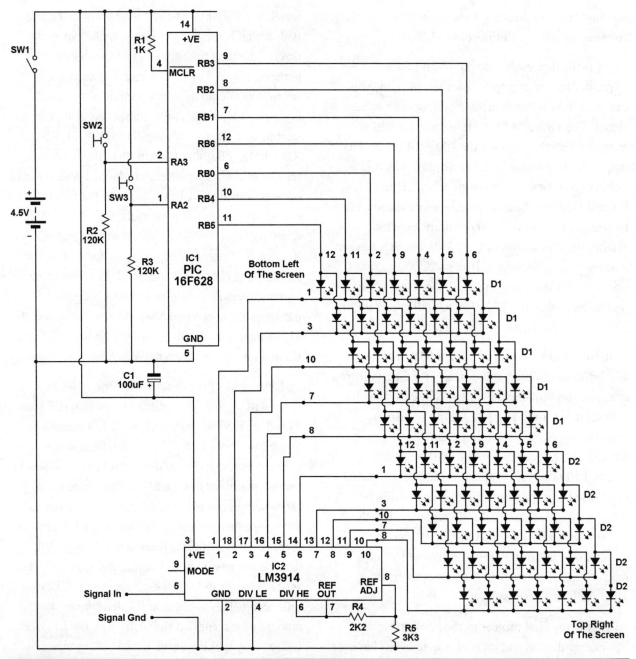

Figure 17-3 The circuit diagram for the digital oscilloscope screen

11mA per LED. This means that even if two LEDs are lit, the total current is still acceptable for each output of IC1, which can drive up to 25mA.

R5 of IC2 is chosen to configure the full-scale deflection of the input voltage, and therefore the Y axis. This design includes the capability to alter the value of this resistor easily for experimental purposes.

R5 is calculated as follows:

$$R5 = ((V_{out} / 1.25) - 1) \times R4$$

Or, by altering the formula:

$$V_{out} = 1.25 (1 + (R5/R4))$$

So, to display a 0–3V input voltage range on the display, resistor R5 should be around 3.3KΩ when we use a 2.2KΩ resistor for R4.

The input voltage to the circuit is fed into pin 5 of IC2 and uses a common ground connection. C1 is included as a decoupling capacitor.

Parts List

> **NOTE** The Supplier and Part Number column of the following table lists specific parts that I used in this project. Refer to the appendix for additional details about acquiring your parts.

The parts you'll need for the digital oscilloscope screen project are listed in the following table.

PARTS LIST			
Code	Quantity	Description	Supplier and Part Number
IC1	1	PIC16F628-04/P Microcontroller	RS Components 379-2869 (Mfr: MicrochipTechnology Inc. PIC16F628-04/P)
IC2	1	LM3914N bar-graph driver IC	RS Components 534-2977
R1	1	1KΩ 0.5W ±5% tolerance carbon film resistor	—
R2, R3	2	120KΩ 0.5W ±5% tolerance carbon film resistor	—
R4*	1	2.2KΩ 0.5W ±5% tolerance carbon film resistor	—
R5*	1	3.3KΩ or 6.8KΩ 0.5W ±5% tolerance carbon film resistor (see text)	—
D1, D2	2	Common cathode red dot-matrix display V_F (typical) = 2V, I_F (typical) = 20mA	RS Components 451-6644 (Mfr: Kingbright TC07-11EWA)
C1	1	100μF 10V radial electrolytic capacitor	—
SW1	1	Single-pole panel mount toggle switch, 2A rated	RS Components 710-9674
SW2/SW3	2	0.24" × 0.24" (6mm × 6mm) tactile momentary push-to-make switch 0.67" high (17mm), 50mA rated	RS Components 479-1463 (pack of 20)
Hardware	1	Stripboard, 0.1" (2.54mm) hole pitch, 37 holes wide by 24 tracks high	—
Hardware	2	18-pin DIL socket	—
Hardware	2	20-way turned pin SIL socket strip	RS Components 267-7400 (pack of 5)
Hardware	2	AA battery holder (three AA batteries)	Maplin YR61R
Hardware	2	PP3 battery clip and lead	RS Components 489-021 (pack of 5)
Hardware	6	AA battery (1.5V)	—
Hardware	—	555 flasher board from Chapter 4 and interconnecting wires (see text)	—
* Note: If you use dot-matrix displays that have different V_F and I_F values to those that are in the parts list, then you may need to alter the resistance and wattage values of these resistors. Please refer to the calculations outlined earlier in this chapter (and the relevant manufacturers' datasheet for IC2) that explain how to do this. You also need to make sure that the maximum reverse voltage allowable for the LED dot-matrix display that you use is higher than the 4.5V supply voltage of the circuit. Also refer to Chapter 12, which discusses the current capabilities of the PIC16F628-04/P microcontroller.			

Stripboard Layout

The stripboard layout for this circuit is shown in Figure 17-4. As you can see, you need to make 46 track cuts in the stripboard, these are shown as white rectangular blocks.

How to Build and Test the Board

NOTE Please refer to Chapter 1 for soldering tips and techniques and for generic stripboard building guidelines.

Build the stripboard by carefully following the diagram shown in Figure 17-4. Note that the two dot-matrix displays are not soldered directly to the board; instead, they fit into four 6-way turned pin SIL sockets, which means that you need to cut two 20-way SIL sockets to size. Also note that you need to make some of the interconnections required around D1 and D2 on the copper side of the board using *insulated* copper wire. This includes pin 7 of D1, which is connected to the track connected to pin 16 of IC2. In addition to this, the following pins of D1 and D2 need to be soldered together on the track side of the board, as shown in Figure 17-5:

- D1 pin 2 to D2 pin 2
- D1 pin 4 to D2 pin 4
- D1 pin 5 to D2 pin 5
- D1 pin 6 to D2 pin 6
- D1 pin 9 to D2 pin 9
- D1 pin 11 to D2 pin 11
- D1 pin 12 to D2 pin 12

You also need to cut some more SIL sockets to size, which will then become connections for the voltage input (2-way SIL) and resistor R5 (two 1-way SIL). Once you have built the stripboard, it

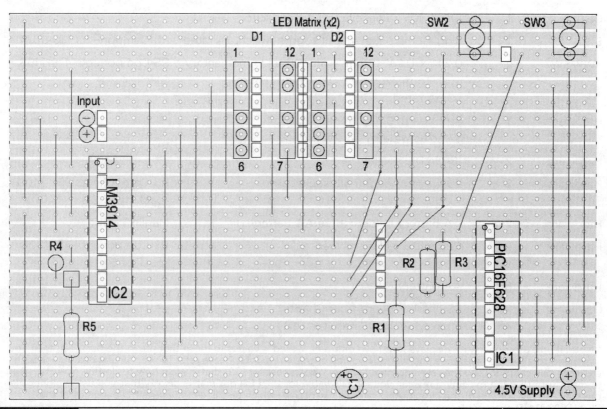

Figure 17-4 The stripboard layout for the digital oscilloscope screen

Chapter 17 ■ Using a Dot-Matrix Display to Show a Waveform **193**

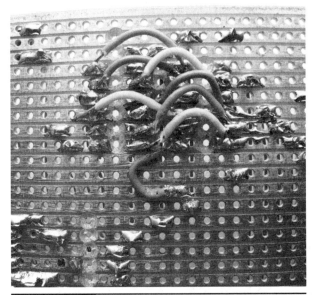

Figure 17-5 Close-up of the track side showing the display connections

should look like the photographs shown in Figures 17-6 and 17-7.

TIP Make sure that you position the two dot-matrix displays in the correct orientation before you fit them into the SIL sockets. The underside of the dot-matrix display that I used is shown in Figure 17-8, and you will notice that there is a raised dimple on one side of the display which denotes the edge that contains pins 7 to 12. This means that the raised dimple of each display is facing towards the right-hand side of the board if you look at Figure 17-7. If you use a different type of dot-matrix display to the one in the parts list it may have different pin-outs to the one that I used, and if this is the case then you will need to modify the stripboard layout to suit your display. Always check the manufacturer's data sheet to confirm the pin configuration of your dot-matrix display.

Figure 17-6 Close-up of the display SIL sockets before the displays are fitted

194 Brilliant LED Projects

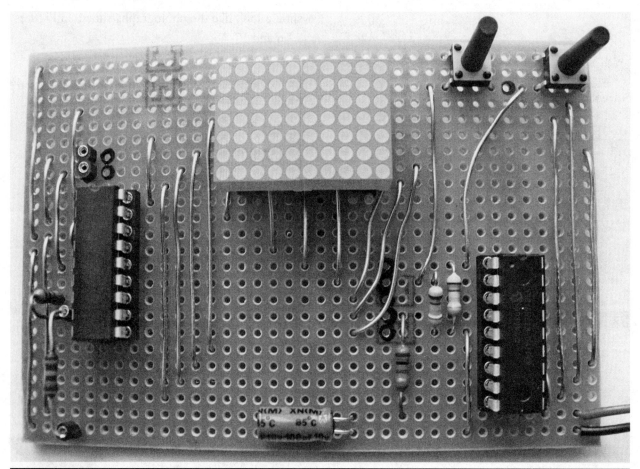

Figure 17-7 The completed stripboard with the two displays fitted

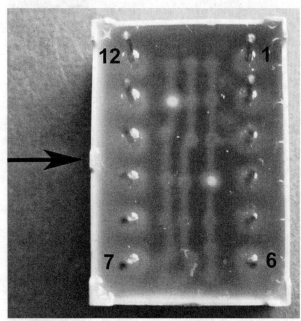

Figure 17-8 The pin configuration of the dot-matrix display that I used (the arrow shows the position of the raised dimple)

Before you fit IC1 and IC2 into their DIL sockets, you can apply the 4.5V battery to the board and test that each of the 70 LEDs in the display matrix works correctly by following the circuit diagram and using a wire link and a 1KΩ resistor. Check the layout of your board carefully, and make sure that there are no solder splashes joining tracks together before you start this procedure; otherwise you could damage the LED displays. Figure 17-9 shows how to illuminate and test the LED that is closest to the "Bottom Left" text on the circuit diagram (shown in Figure 17-3).

To do this, you need to wire the 1KΩ resistor between pin 1 (output 1) and pin 2 (ground) of the DIL socket of IC2, which feeds a current-limited negative voltage to the dot-matrix display, and you also need to connect a wire link between pin 14 (positive) and pin 11 of the DIL socket of IC1. You

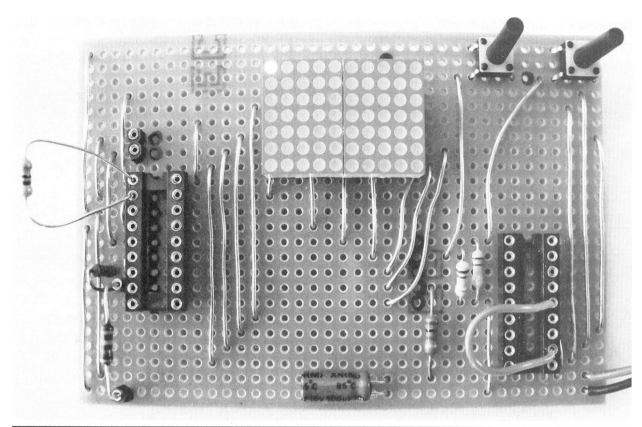

Figure 17-9 Use a resistor and a wire link to test each of the LEDs.

can continue to test each of the LEDs in the first column by keeping one end of the wire link of IC1 connected to pin 14 and moving it along to pins 11, 10, 6, 12, 7, 8, and 9 of IC1 in turn. After you have tested the first column, connect the 1KΩ resistor between pin 18 (output 2) and pin 2 (ground) of IC2's DIL socket, and then repeat the wire link procedure on IC1. Continue this process until you have tested all 70 LEDs. It is important that you always use the 1KΩ resistor when linking pin 2 (ground) to each of the output pins of IC2's DIL socket; otherwise you will damage the LEDs in the dot-matrix display. Only one LED should be illuminated at any one time; if no LED illuminates or if more than one illuminate, then you need to check the board over for errors or solder splashes that may be linking copper tracks together. When you have completed the testing procedure, leave the wire link and resistor in place and then remove the battery from the board; this will let the capacitor discharge through one of the LEDs. Once this has been done you can remove the wire link and resistor, and then move on to programming IC1.

The PIC Microcontroller Program

Now you need to program IC1 and install it on the stripboard. You can download the assembly program and Hex file from the McGraw-Hill website, at www.mhprofessional.com/computingdownload. You can then program IC1 with the Hex file called **LED Oscilloscope Screen.hex** as explained in Chapter 12.

The Assembly Program

The assembly program that creates the operation of the sequencer circuit for IC1 is called **LED Oscilloscope Screen.asm**. It includes notes that explain its operation in more detail, but basically the seven Port B outputs are made to activate high in sequence, one at a time, with pauses in between, in a never-ending loop. An extract of this routine is provided here:

```
DISPLAY  bsf  PORTB,5
         call PAUSE
         clrf PORTB
         bsf  PORTB,4
         call PAUSE
         clrf PORTB
         bsf  PORTB,0
         call PAUSE
         clrf PORTB
         bsf  PORTB,6
         call PAUSE
         clrf PORTB
         bsf  PORTB,1
         call PAUSE
         clrf PORTB
         bsf  PORTB,2
         call PAUSE
         clrf PORTB
         bsf  PORTB,3
         call GETKEY
         goto DISPLAY
```

The two Port A inputs are monitoring the two switches, SW2 and SW3. Pressing these buttons increases or decreases the "speed" variable, which alters the speed at which the sequencer circuit operates. This means that you can adjust the display scanning speed to stabilize the digital signal input on the display.

The Hex File

The file for the hex code that you need to download to program into IC1 is called **LED Oscilloscope Screen.hex**. A complete version of the code is shown here:

```
:020000000528D1
:08000800052807309F00850167
:10001000860183160C308500030860080308100018
:1000200083123230A200A30186162920860106160B
:100030002920860106142920860106172920860119
:10004000861429208601061529208601861534206C
:1000500014283230A0002208A100A00B2D28323035
:10006000A000A10B2D2808000508003A03194A2812
:100070002318080085193120A0519A20B4028472059
:10008000A21B44202314080000130A20008007F3086
:08009000A2000800A3011428DE
:02400E00303F41
:00000001FF
```

Give Me a Wave

Now for the fun part—viewing some waveforms. First of all, making sure that there is no voltage applied to the board, insert IC1 and IC2 into their DIL sockets and fit a 3.3KΩ resistor to the SIL socket position for R5. Next, apply the 4.5V supply to the board. Turn the board so that the narrow edge of the board is nearest to you and SW2 and SW3 are near the top of the board; this is the orientation in which you should view the display from now on. You should immediately see the top one or two lines of the display illuminated, and scanning at a very fast rate, similar to what is shown in Figure 17-10.

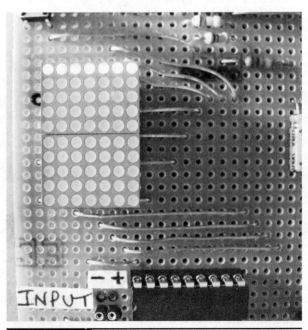

Figure 17-10 The display with no input applied

Now press SW2 a number of times. This should start to reduce the speed of the sequencer circuit and the scanning speed. Then press SW3 a number of times. This should increase the speed until the sequence is moving so fast that you can no longer perceive the scanning and the top two rows appear to be illuminated permanently. This is a POV illusion—the sequencer is operating at a fast enough rate to make it look as though 14 LEDs are permanently lit, when in fact only one or two LEDs are illuminated at the same time. The circuitry around IC2 is so sensitive that even touching the copper tracks of the board can start to produce some images on the display, but you should try to avoid doing this.

Using the 555 LED flasher stripboard that you built in Chapter 4, insert two 18KΩ resistors into the R1 and R2 sockets, and insert a 1µF (min. 10 volt rated) electrolytic capacitor into the C1 socket. Make sure that the variable resistor is wound fully clockwise and then apply to the 555 LED flasher board a separate 4.5V supply using three AA batteries (instead of a 6V supply, as used in Chapter 4). The LED on the flasher stripboard should start to flash at a very fast rate, oscillating at around 25 Hz.

Figure 17-11 shows how to interconnect the 555 flasher board to the LED oscilloscope display. Notice that I have soldered some additional single SIL sockets to the 555 LED flasher board to make it easier to connect the two boards together; these additional pins are connected to ground and pin 6 of the 555 timer.

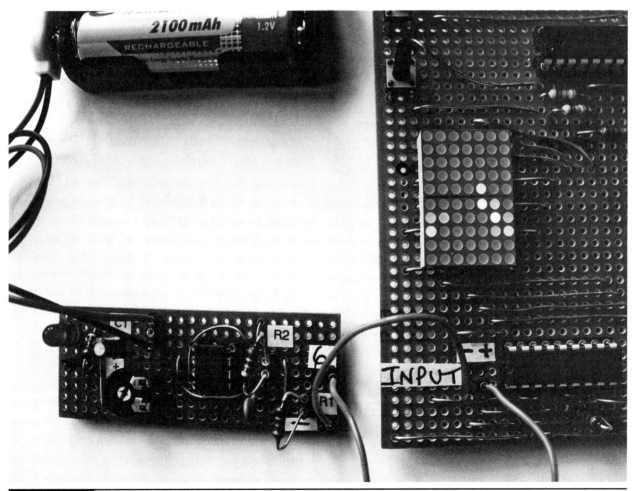

Figure 17-11 Connecting the 555 timer board to the "scope"

Connect the negative input on the oscilloscope screen to the ground connection on the 555 timer board, and connect the positive input on the oscilloscope screen to pin 6 of the 555 timer board. In astable mode, pin 6 of a 555 timer produces a sawtooth-type waveform, which (hopefully) you should be able to see on the display of your scope. At first you might just see some very fast blurry images, but by increasing or decreasing the sequencer speed—by pressing SW2 or SW3—you should be able to "tune in" the image of the waveform on the display, and it should look something like the image shown in Figure 17-12. Note that you have to press and release the speed buttons to increment the speed up or down a notch; keeping your finger on either button will not automatically alter the speed of the display. By playing around with the time base speed, you should find that increasing the speed has the effect of zooming into the waveform image at certain timing points.

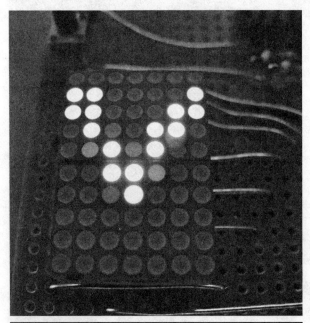

Figure 17-12 Sawtooth waveform from pin 6 of the 555 timer

TIP The test points that you are connecting to on the 555 timer board have a peak output voltage of around 2–3 volts, so you should be able to see them clearly on the display.

You can experiment with the display board by changing the value of R5 so that the scale of the voltage shown on the display is increased or decreased. If you find that the waveform is clipped at the top of the display, try increasing the resistor value of R5 to 6.8KΩ.

If you now connect the + signal input of the oscilloscope board to pin 3 of the 555 flasher board, you should be able to see a square wave output on the display similar to that shown in Figure 17-13.

You will notice that the display doesn't show a textbook square wave; it only looks as if a few LEDs are lit at the top of the display. This is because the square wave switches from a high to low state, and a low to high state, at a very fast rate which means that the vertical changes are not able to be seen on the display. This results in you just seeing the high parts of the waveform. If you use the two speed switches to alter the speed of the display, you will eventually see a line of LEDs lit at the top of the display with a gap in between them; the gap is actually the low state in between two high parts of the waveform.

Experiment with the two circuits and the speed of the 555 timer to see if you can see the images clearly on the scope display.

You can alter the value of R5 on the scope board to match the maximum output voltage of the circuit being tested. This requires that you calculate the resistor values using the formulas presented earlier in the chapter. I do not recommend that you test any waveform circuits that have a DC voltage higher than 5V with the current circuit design.

Figure 17-13 Square wave from pin 3 of the 555 timer

CAUTION The input to the LED oscilloscope screen should always be a positive DC input. Do not attempt to measure an AC voltage in its current format because the LM3914 will not accept a negative or reverse voltage, and trying this could damage the circuit.

You might also want to consider wiring a 0.25W-rated 10KΩ variable resistor in place of R5 so that you can vary the upper limit of the display voltage. Reducing the resistance of R5 reduces the upper input voltage limit that can be displayed.

CAUTION Do not input a higher voltage into the scope board than is set by R5. For example, if R5 is configured to accept a maximum voltage of 3V, do not input a waveform into the scope screen that has a 4.5V peak voltage; this could damage IC2, and you wouldn't be able to see the full waveform anyway.

Other Ideas

Experimenting with the LED digital oscilloscope screen circuit should help you to think of some future circuit designs and applications for this board. In my original design in 1994, I also added some amplifier circuitry to magnify the input voltage signal, and some additional circuitry to measure the frequency of the waveform and display it on two seven-segment displays. These additional circuit modules are beyond the scope of this book, but you can see that by adding some additional circuit modules, you could utilize the LED oscilloscope screen in a much larger project.

One other application that springs to mind is that you could build an amplifier circuit connected to a microphone, which would allow you to convert the sound of music or a noise into a suitable voltage signal that could be viewed on the display screen. I have not tried this, but I imagine that you could probably use such a circuit to produce some interesting sound-to-light effects on the display. In fact, you could create an interesting display effect by simply connecting pin 6 of the 555 LED flasher circuit to show the sawtooth waveform and then adjusting the timing slightly so that the waveform moves along the display like a roller coaster.

CHAPTER 18

Light-Dependent LEDs: Experimental Low-Res Shadow Camera

THE PROJECT PRESENTED IN THIS CHAPTER STARTED out as a bit of an experiment. I wanted to see if I could emulate the operation of a digital camera sensor. As you'll see, the end result is not exactly how the sensor in a digital camera works (which is why I've labeled the project "experimental"), but it does produce a really interesting light effect. The project comprises a light-sensitive sensor array, the output of which changes depending on the amount of light that it "sees." This image is then replicated on a display "screen," which is created using a matrix of LEDs. Each LED in the "screen" is switched on and off in turn at a very fast rate creating a POV effect, which fools the eye into seeing a continuous display of light. The experimental low-res shadow camera stripboard is shown in Figure 18-1.

As I'll discuss at the end of the chapter, this project also provides some interesting perspective on how we have come to take for granted the miniaturization of the electronics that we use in our daily lives.

Project 17
Experimental Low-Res Shadow Camera

This project uses a PIC microcontroller to generate a POV effect; if you have not already done so it is worth reading Chapter 12, which explains the key specifications of this device and how to program it.

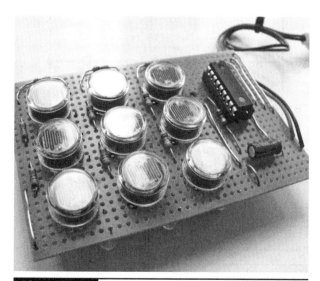

Figure 18-1　The experimental low-res shadow camera

PROJECT SPECIFICATIONS
■ The light-sensitive sensor array comprises nine pixels in a 3 × 3 format.
■ The LED "screen" is a nine-pixel display (3 × 3 format).
■ The LED "screen" is created using POV.
■ The circuitry is microcontroller driven.
■ The supply voltage is 4.5 volts.

201

202 Brilliant LED Projects

It is also recommended that you read Chapter 15, which explains how POV works.

How the Circuit Works

The circuit diagram for the experimental low-res shadow camera is shown in Figure 18-2.

The circuit is powered by three AA batteries, which produce a 4.5V supply. This drives IC1, which is a PIC16F628-04/P microcontroller. The circuit diagram shows a power switch (SW1), which is a toggle switch; however, because this was an experiment, I did not use a power switch in my prototype as you will see shortly; however, you may decide to use one in your project. All of the I/O ports of IC1 are configured as outputs in the software, and nine IC1 ports are used, RB1 through RB7 and RA2 and RA3. Each output is fed through a current-limiting resistor (R2–R10),

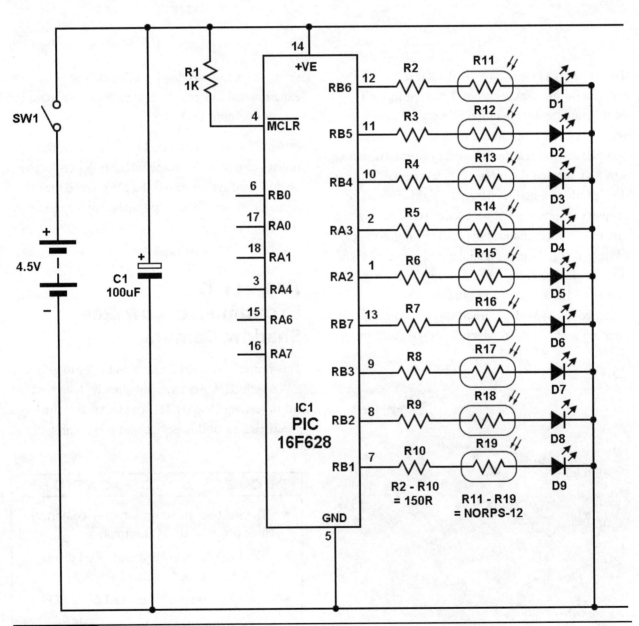

Figure 18-2 The circuit diagram for the experimental low-res shadow camera

which in turn is connected to a light-dependent resistor (R11–R19) before being fed into an LED (D1–D9). This configuration is identical for all nine outputs.

A light-dependent resistor (LDR) is a resistor whose resistance alters depending on how much light is presented to it. A common use for this component is in a security night light that switches on at dusk. In ambient daylight conditions, the LDR used in this circuit has a resistance of around 2KΩ; in the dark, the resistance increases to 500KΩ+; and in really bright light, the resistance drops to around 100Ω.

Ignoring IC1 for now, the principle of the circuit operation is that the brightness of each LED is dependent on the amount of light that shines on its associated LDR. For example, in ambient daylight conditions, the total series resistance feeding into an LED is approximately 2.15KΩ (150Ω series resistor + 2KΩ LDR). Shining a bright light onto the LDR causes this series resistance to decrease to around 250Ω (150Ω + 100Ω). This means that if you shine a bright light onto an LDR, its associated LED illuminates at a reasonable brightness level. If the light level falls, the LED shines less brightly. By positioning the nine LDRs and LEDs into a three-by-three format, you can start to create a light-sensitive array, which means that the light input to the array can be replicated on an LED display.

NOTE There are LDRs available that are smaller than the type used in this project. I chose the version used in this project because it can dissipate up to 250mW, which is enough for the operation of this circuit. The smaller-sized LDRs have a much reduced current/wattage capability and therefore may not be suitable in this application.

You could in fact build this circuit without any driver circuitry, doing away with IC1 altogether, but this would mean that all nine LEDs would be powered at the same time and the current consumption of the circuit could be over 100mA, depending on the light level presented to the array. The reason for including IC1 is to produce a sequencing effect, so that only one LDR/LED is powered at any one time. This sequencer circuit is similar to the 555/4017 LED sequencer circuit presented in Chapter 8. In this circuit, however, the operation is performed at a very fast rate, which means that all nine LEDs look as though they are lit at the same time. This helps to reduce the overall current consumption of the circuit, and it also demonstrates another way of creating a POV effect using a constantly changing voltage signal. C1 is included in the circuit as a decoupling capacitor.

Parts List

NOTE The Supplier and Part Number column of the following table lists specific parts that I used in this project. Refer to the appendix for additional details about acquiring your parts.

The parts you'll need for the experimental low-res shadow camera project are shown in the table that follows.

PARTS LIST

Code	Quantity	Description	Supplier and Part Number
IC1	1	PIC16F628-04/P Microcontroller	RS Components 379-2869 (Mfr: MicrochipTechnology Inc. PIC16F628-04/P)
R1	1	1KΩ 0.5W ±5% tolerance carbon film resistor	—
R2–R10*	9	150Ω 0.5W ±5% tolerance carbon film resistor	—
R11–R19	9	Light-dependent resistor Bright resistance = approx. 100Ω Dark resistance = approx. 500KΩ+ Wattage = 250mW	RS Components 651-507 (Mfr: Silonex NORPS-12)
D1–D9	9	5mm yellow LED V_F (typical) = 2.1V, I_F (typical) = 20mA	RS Components 228-6010 (pack of 5)
C1	1	100μF 10V radial electrolytic capacitor	—
SW1	1	Single-pole panel mount toggle switch, 2A rated (optional)	RS Components 710-9674
Hardware	1	Stripboard, 0.1" (2.54mm) hole pitch, 37 holes wide by 24 tracks high	—
Hardware	1	18-pin DIL socket	—
Hardware	1	AA battery holder (three AA batteries)	Maplin YR61R
Hardware	1	PP3 battery clip and lead	RS Components 489-021 (pack of 5)
Hardware	3	AA battery (1.5V)	—
Hardware	1	Piece of thick black card (A4 size), piece of tracing paper (A4)	Office supply/stationary store
Hardware	—	Double-sided adhesive tape, black insulation tape, flashlight	—

* Note: If you use LEDs that have different V_F and I_F values to those that are in the parts list, then you may need to alter the resistance and wattage values of these LED series resistors. Please refer to Chapter 2, which explains how to do this. Also remember that the LDRs (R11–R19) have an effect on the amount of current flowing through the LEDs. I calculated the LED series resistor values by assuming that the LDR has zero resistance, and I used a value of 16mA for I_F in the formula. This means that the maximum current that can flow through the LED is 16mA, because it is unlikely that the LDR will ever have zero resistance. You also need to consider the maximum power rating of the LDRs in your calculations.

Stripboard Layout

The stripboard layout for this project is shown in Figure 18-3. As you can see, you need to prepare the board by creating the 23 track cuts, which are shown as white rectangular blocks in the layout diagram.

How to Build and Test the Board

NOTE Please refer to Chapter 1 for soldering tips and techniques and for generic stripboard building guidelines.

Build the stripboard by carefully following the diagram shown in Figure 18-3. However, *do not* fit the nine LEDs (D1–D9) at this stage. You will fit them to the track side of the board in the next

Chapter 18 ■ Light-Dependent LEDs: Experimental Low-Res Shadow Camera

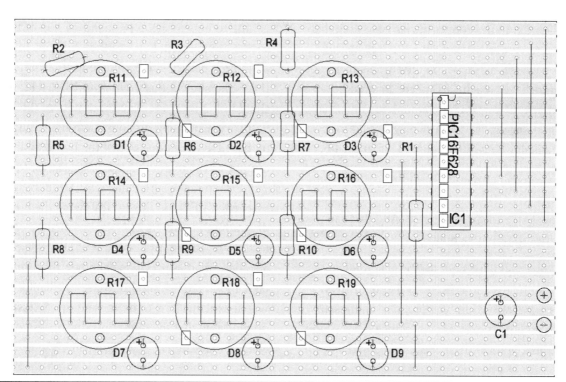

Figure 18-3 The stripboard layout for the experimental low-res shadow camera. Note that the zigzag lines in R11–R19 represents the internal structure of the LDR, they are not wire links.

stage. After you complete this first stage of the stripboard construction, it should look like the stripboard shown in Figure 18-4 (do not fit IC1 into the DIL socket yet).

Now turn the board over so that the copper track side is showing. Solder the nine LEDs in place. You need to make sure that each LED stands about 5mm above the board so that you can easily solder

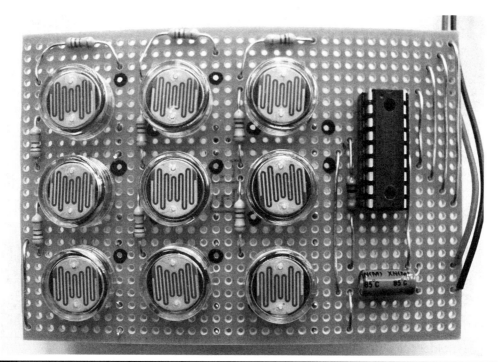

Figure 18-4 The camera side of the stripboard

the LED leads to the copper tracks. Also note that you need to carefully bend the LED legs apart slightly because, as shown on the stripboard layout (see Figure 18-3), each LED needs to straddle a copper track. Once you have completed this final stage of soldering, the LED display should look like the display shown in Figures 18-5 and 18-6.

After you solder the LEDs into place, you can cut down the LED leads that protrude through the component side of the stripboard. And that's it—the electronics are now built.

Now that you have built the circuit, you can perform some tests to make sure that the LEDs work as expected before you program and install IC1 into its DIL socket. If you apply a 4.5V supply to the board, you can first check that power is getting to pins 4 (+), 5 (−), and 14 (+) of the DIL socket as expected. You can then link pin 14 to pin 12 to provide a positive voltage to R2, R11, and D1. Next, shine a flashlight onto LDR R11. This should illuminate D1. The brightness of the LED will vary depending on how much light you shine on the LDR. Continue to test the board by linking out pin 14 (+) to pins 12, 11, 10, 2, 1, 13, 9, 8, and 7 in turn to make sure that each LDR activates its associated LED; do this by following the circuit diagram in Figure 18-2. A similar method of testing the output pins of the IC's DIL socket is also described in previous chapters; for example, take a look the testing procedure in Chapter 12 which explains how to use a wire link to test each output (note that the output pins to be tested in this chapter are different from those in Chapter 12).

Figure 18-5 Solder the LEDs to the track side.

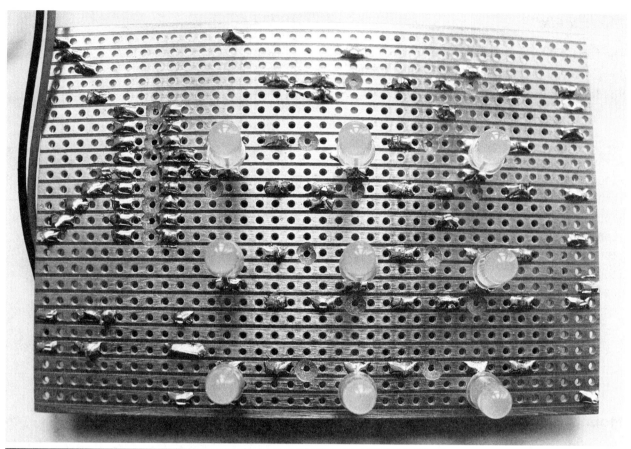

Figure 18-6 The completed three-by-three LED matrix display

The PIC Microcontroller Program

Now you need to program IC1 and install it on the stripboard. You can download the assembly program and Hex file from the McGraw-Hill website, at www.mhprofessional.com/computingdownload. You can then program IC1 with the Hex file called **LED Shadow Camera.hex**, as explained in Chapter 12.

The Assembly Program

The assembly program for this project is called **LED Shadow Camera.asm**. The program is fairly basic and, as usual, includes notes that help to explain its operation.

Essentially, the program works by switching on each of IC1's outputs one at a time, starting with RB6. After an output is switched on, there is a slight delay before it is switched off. The next output is then switched on and off in the same manner, and so on in a never-ending sequence. This is performed at a very fast rate and creates a POV effect, which means that it looks like all the LEDs are illuminated at the same time. The following is an extract of the program showing part of this routine that controls the first three LDRs:

```
CAMERA: bsf PORTB,6    ;activate 1st LDR
        call PAUSE     ;delay
        bcf PORTB,6    ;de-activate 1st LDR

        bsf PORTB,5    ;activate 2nd LDR
        call PAUSE     ;delay
        bcf PORTB,5    ;de-activate 2nd LDR

        bsf PORTB,4    ;activate 3rd LDR
        call PAUSE     ;delay
        bcf PORTB,4    ;de-activate 3rd LDR
```

The Hex File

The file for the hex code that you need to download and program into IC1 is called **LED Shadow Camera.hex**. The complete hex code listing is presented here:

```
:020000000528D1
:08000800052807309F00850167
:10001000860183160030850000308600803081 0024
:100020008312850186010617 2F20061386162F20BE
:100030008612061 62F20061285152F20851105150C
:100040002F20051186172F20861386152F20861145
:1000500006152F20061186142F2086101328043031
:10006000A0008B010B1D32280B11A00B32280800B9
:02400E00303F41
:00000001FF
```

Once you have programmed IC1, you can fit it into the DIL socket on the completed stripboard.

Mount the Board

You might decide to find a suitable enclosure in which to house the battery holder and stripboard. Because I was experimenting, I never got around to housing my stripboard in an enclosure. However, I did need to make a light shield so that I could test the board. I found a really thick (A4 size) piece of black card (which I had previously purchased from an office store) and punched nine 5mm holes in it to match the LED display layout. First of all, I overlaid the LED side of the stripboard with a piece of tracing paper, and marked onto it the center position of all nine LEDs. I then used this template to mark out the LED positions onto the black card before punching the nine holes in the card. The final result is shown in Figure 18-7.

I then aligned the LEDs with the holes in the card and gently pushed the stripboard so that the LEDs protruded through the card, as shown in Figure 18-8. You should find that if the card is thick enough, the board will sit fairly securely in

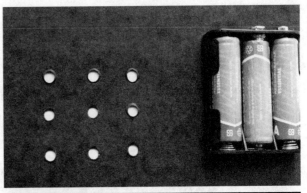

Figure 18-7 Punch nine holes in the black card.

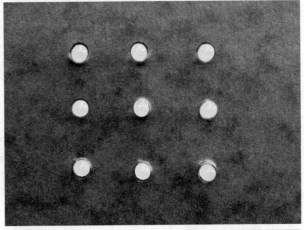

Figure 18-8 The LEDs protrude through the holes in the card to produce a display.

place. I next flipped the board over and affixed the battery holder to the card by using double-sided adhesive tape so that it would sit neatly next to the stripboard, as shown in Figure 18-9.

Finally, I cut a smaller piece of black card to create a mask for the light sensor array. This mask is to help stop the light from the flashlight that is shone onto the light sensor from permeating through the holes in the stripboard and shining through the LEDs, which can spoil the effect. I cut the card carefully by using a craft knife and then rounded the corners by affixing thin strips of black insulation tape to the back of the mask, so that they fit snugly around each of the LDRs. The completed camera sensor mask is shown in Figure 18-10.

Chapter 18 ■ Light-Dependent LEDs: Experimental Low-Res Shadow Camera   209

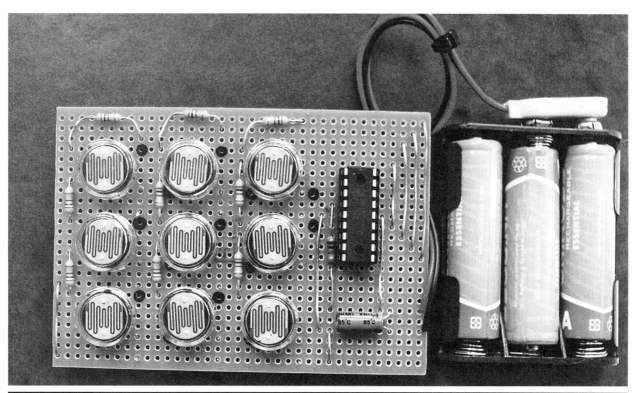

Figure 18-9 The stripboard sits neatly next to the battery holder on the camera side.

Figure 18-10 The sensor array mask

If you have followed my example, you can test how well the mask works by shining a bright light onto the LDR array (without power applied to the circuit) and seeing if you can detect any light shining through any of the LEDs. Don't be concerned if a small amount of light appears through the LEDs, because this will not spoil the overall effect.

Seeing a Shadow

Once you have completed construction, you can connect the battery lead to the battery holder and then hold the card (or the enclosure) so that you are looking at the LED display. Depending on the ambient light conditions, you should see either a dimly lit LED display or a display that does not look as though it is lit at all. For best results, experiment with this circuit in the dark. If you are in a dark room, the display will be off; if you then shine your flashlight across the sensor side of the card, you should see this light beam being replicated on the LED display.

You can create some interesting effects with this device. For example, shine the light on the sensor array so that the complete display is illuminated, and then wave your finger or a pencil in between the light source and the sensor array—you will see this image being replicated on the LED display in the form of a shadow. It is a bit surreal if you use your finger because you actually feel as if you are seeing a shadow of your finger on the display. Also, try moving the light source around the sensor array so that the bright light from the flashlight and the darker areas are replicated on the display, as shown in the example in Figure 18-11.

Further Ideas

Have a bit of fun playing around with this project and you might generate some ideas for how you can utilize this circuit in future project designs. One idea is to make more of these circuits and position them together so that a larger object can produce a shadow on the screen. Of course, depending on how many circuits you incorporate into your design, this could be quite an expensive method of producing this effect because of the price of each LDR. If you decide to do this, you might also consider building a larger, single LED display on a separate piece of stripboard and mounting it remotely from the sensor arrays using interconnecting cables.

As I mentioned in the chapter introduction, this project also provides some interesting perspective on how we have come to take for granted the miniaturization of electronics in our daily lives. Think about how we would create a shadow camera like this with a 1-megapixel resolution (that is, 1 million pixels). This would require a matrix of 1000 $\times$ 1000 LDRs. If we were to replicate the same stripboard shape to create this type of display, we would need to build 111,556 stripboard layouts, to form a 334 $\times$ 334 matrix. The approximate size of this layout would be a frightening 105 feet (32 meters) wide by 70 feet (21.3 meters) high! Think about that the next time you take a photograph using your digital camera.

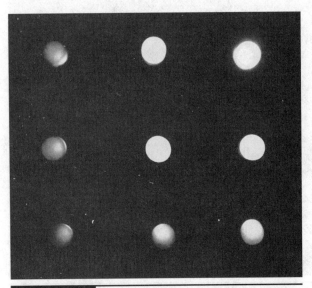

Figure 18-11 Creating shadows on the "screen"

CHAPTER 19

Creating a POV Effect in Mid-Air: Groovy Light Stick

How would you like to make an exciting light display using just five red LEDs? If you are tempted, read on and see how you can build a groovy light stick like the one shown in Figure 19-1. This project produces an interesting POV effect, which is described in Chapter 15, by using a single row of five quickly flashing LEDs to create moving light patterns in the air. The circuit uses a microcontroller to store the light pattern image data, and to switch the five LEDs in a fast moving sequence to generate the effect required.

PROJECT SPECIFICATIONS

- Five red LEDs create a POV light display.
- The circuitry is microcontroller driven.
- The display speed is adjustable using a single push button.
- The POV effect can be seen by waving or rotating the "stick."
- The supply voltage is 3 volts.
- The microcontroller is programmed to incorporate a low current snooze facility, to save battery power when the unit is not in use, and also means that a power switch is not required.

Figure 19-1 The groovy light stick

Project 18
Groovy Light Stick

This project is really simple to build and uses POV to create some interesting light effects. It makes five LEDs flash in a particular sequence that leaves unique patterns of light "hanging in the air" when you wave or rotate the unit.

How the Circuit Works

The circuit diagram for the groovy light stick is shown in Figure 19-2. As you can see, there is not much to the circuit. The bulk of the work is performed by the software programmed into the PIC Microcontroller.

The circuit is powered via two AAA batteries that provide a 3V supply to power the PIC16F628 microcontroller (IC1).

Only six of IC1's I/O ports are used in this project. B1 to B5 are configured as outputs and drive five LEDs (D1 to D5) via series resistors R2 to R6. Port B0 is configured as an input and is normally held low via R9; if push button switch

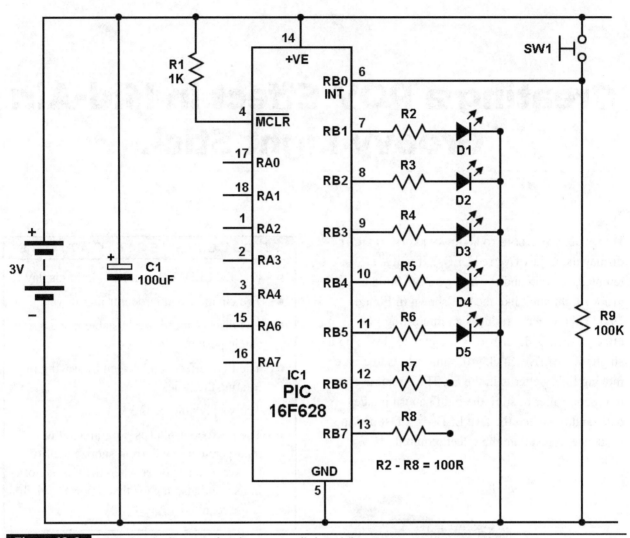

Figure 19-2 The circuit diagram for the groovy light stick

SW1 is pressed, Port B0 is taken high, and this is monitored via the software. The other ports, Ports B6 and B7, are just connected to R7 and R8 and were included in case you decide to modify the circuit and software to include two additional outputs to create a display with seven LEDs rather than five. You can exclude R7 and R8 if you want to.

Resistor R1 provides power to pin 4 as usual to activate IC1. You will also notice that there is no power switch in the circuit; this is because there is a sleep facility written into the software, as explained later in the chapter. C1 is a decoupling capacitor that is included to smooth the supply to the circuit.

Parts List

NOTE The Supplier and Part Number column of the following table lists specific parts that I used in this project. Refer to the appendix for additional details about acquiring your parts.

The parts you'll need for the groovy light stick project are listed in the following table.

Chapter 19 ▪ Creating a POV Effect in Mid-Air: Groovy Light Stick

PARTS LIST

Code	Quantity	Description	Supplier and Part Number
IC1	1	PIC16F628-04/P Microcontroller	RS Components 379-2869 (Mfr: MicrochipTechnology Inc. PIC16F628-04/P)
R1	1	1KΩ 0.5W ±5% tolerance carbon film resistor	—
R2-R8*	7	100Ω 0.5W ±5% tolerance carbon film resistor	—
R9	1	100KΩ 0.5W ±5% tolerance carbon film resistor	—
C1	1	100μF 10V radial electrolytic capacitor	—
D1-D5	5	5mm red ultra-bright LED V_F (typical) = 1.85V, I_F (typical) = 20mA	RS Components 564-009
SW1	1	Single-pole normally open panel mount switch (100mA)	RS Components 133-6502
Hardware	1	Stripboard, 0.1" (2.54mm) hole pitch, 25 holes wide by 9 tracks high	—
Hardware	1	18-pin DIL socket	—
Hardware	1	Ten-way, single-row pin header (cut to size, per the text)	ESR Electronic Components 111-110
Hardware	1	AAA battery holder (two AAA batteries)	RS Components 512-3552
Hardware	2	AAA battery (1.5V)	—
Hardware	1	Small, narrow enclosure, approx. 4.88" (124mm) long × 1.3" (33mm) wide × 1.18" (30mm) deep	Maplin FT31
Hardware	5	5mm LED clips	—
Hardware	—	Cable ties, cable tie bases, multicolor interconnecting cables, double-sided adhesive tape	—

* Note: If you use LEDs that have different V_F and I_F values to those that are in the parts list, then you may need to alter the resistance and wattage values of these LED series resistors. Please refer to Chapter 2, which explains how to do this, and also consider the maximum current sourcing capabilities of IC1, which is outlined in Chapter 12.

Stripboard Layout

The stripboard layout for the groovy light stick is shown in Figure 19-3. It is very similar (but not the same) as the layout for the LED light sword project described in Chapter 13. Note that you need to make 16 track cuts before you solder the components into place; these are shown as white rectangular blocks on the layout diagram.

How to Build the Board

NOTE Please refer to Chapter 1 for soldering tips and techniques and for generic stripboard building guidelines.

First, to fit the stripboard inside the enclosure, you need to cut the stripboard so that it is slightly smaller (24 holes wide by 9 tracks high) and modify two of its corners, as shown in Figure 19-4. The stripboard specified in the parts list is 25 holes wide by 9 tracks high, so you need to remove only

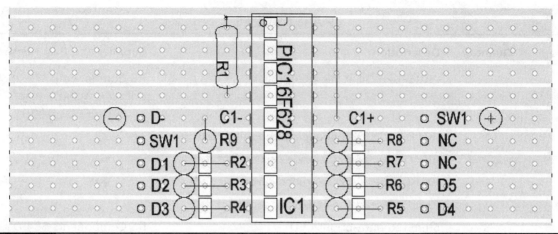

Figure 19-3 Stripboard layout for the groovy light stick

a single row of holes from the width before you modify the two corners.

Build the stripboard layout by closely following the diagram shown in Figure 19-3. You provide the solder points for the battery connections, the five LEDs (D1–D5), and SW1 by soldering two five-pin headers on either side of IC1's DIL socket. Also note that because the stripboard has limited track numbers, you need to fit C1 to the track side of the stripboard and solder it across pins 5 and 14 of IC1, as shown in Figure 19-4. Make sure that the polarity of C1 is correctly oriented, so that the negative lead of the capacitor is facing pin 5 of IC1, before you solder it into place.

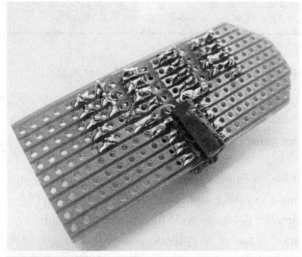

Figure 19-4 Solder C1 to the track side of the board.

The completed board should look like the example shown in Figure 19-5.

Build the LED Display

After you have completed the stripboard, your next task is to build the LED display, which is fitted to the lid of the enclosure. Start by drilling into the lid five equally spaced holes that are large enough to accept the LED clips. Leave a slight gap in between each LED. Then insert the five LEDs into their clips, making sure that the polarity of each of the LED leads is positioned in the same orientation for all five LEDs.

Next, carefully bend over the LED leads and solder all five cathode (−) leads together, and then solder a flying lead to this common cathode connection. Then solder five more flying leads from the anode (+) connections of each LED and secure all six cables together using cable ties. Secure the wiring loom to the back of the lid using a cable tie base. I suggest that you use different color cables for these connections, which makes it easier to identify the connections once you terminate them to the solder points on the stripboard. The finished display should look like the images shown in Figures 19-6 and 19-7.

Figure 19-5 The completed stripboard for the groovy light stick

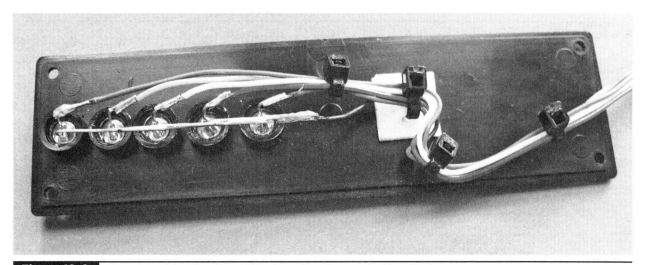

Figure 19-6 How to solder the five LEDs together to create the display

Figure 19-7 The front of the display should look like this.

Complete the Enclosure

You can now prepare the inside of the box by fixing the battery holder into place using double-sided adhesive tape and fitting two cable tie bases, one to attach the LED cables and the other to help to hold the stripboard in place. Then make sure that the stripboard fits neatly in place and mark out a suitable position for the switch and fit this into the enclosure. Once everything fits in place, remove the stripboard and the enclosure should now look like the photograph in Figure 19-8. It

Figure 19-8 Preparing the enclosure

should be noted that the position of the switch is quite important, so read the following text and look at Figure 19-9 before drilling the hole for the switch.

Now attach the LED cables to the cable tie base on the side of the enclosure, making sure that there is enough slack in the cables to be able to open and close the lid. Then, solder all of the

Figure 19-9 Everything is soldered in place.

interconnecting cables to the pin headers on the stripboard. There isn't a lot of room inside the enclosure, so take your time when soldering the interconnecting cables together (it is a bit like performing electronic surgery!). After you solder all of the cables into place, the finished enclosure should look like Figure 19-9. Note how the stripboard is secured to the cable tie base using a cable tie, and also notice how I have managed to position the switch (SW1) so that one of its leads can be soldered directly to the top right-hand pin header (marked SW1 + on Figure 19-3) and to the positive battery lead. These two mechanical features help to stop the stripboard from moving around when it is being waved in the air. The negative battery lead and the common cathode connection for the LEDs are soldered to the top left-hand pin header marked D– on Figure 19-3.

Test the Board

Before you program and install IC1, you can make the usual power checks to ensure that there are 3 volts in the correct polarity across pins 5 and 14 of the DIL socket. You can also check that pin 6 is normally low and then goes high when SW1 is pressed. With 3 volts applied, you can then follow the circuit diagram in Figure 19-2 to test the operation of the five LEDs. You can test the LEDs by connecting the positive voltage from pin 14 of the DIL socket to pins 7 to 11 one at a time using a piece of wire, to make sure that each of the five LEDs illuminates as expected; see Figure 19-10.

When you have proved the operation of the board and you have tested the last LED (D5), remove the batteries from the circuit and let capacitor C1 discharge through the LED for a few seconds before removing the wire link.

Figure 19-10 Pin 14 (+) of IC1's DIL socket is connected to pin 11 and illuminates the last LED (D5).

The PIC Microcontroller Program

Now you need to program IC1 and install it on the stripboard. You can download the assembly program and Hex file from the McGraw-Hill website, at www.mhprofessional.com/computingdownload. You can then program IC1 with the Hex file called **LED Groovy Light Stick.hex**, as explained in Chapter 12.

The Assembly Program

The assembly program for this project is called **LED Groovy Light Stick.asm**. The majority of the program is taken up by a look-up table that contains all of the binary data required to create the images. The software works in a similar manner to that of the backpack illuminator project in Chapter 16, except that instead of multiplexing the display column by column, the changing display data is continuously presented to the single row of LEDs.

There are some variables described in the software that determine the speed of the display. Push button switch SW1, which is connected to input B0, is being monitored by the software, and this allows you to adjust the display speed by pressing the button. The other feature of the software is that after the LEDs have gone through their display sequence five times, the program goes into sleep mode. This means that the microcontroller essentially powers down and draws a very low current (microamps) from the supply until it receives an external trigger to wake it up again. The benefit of this feature is that you do not need to have a separate power switch for this circuit. The reason that SW1 is connected to input Port B0 is so that this input can be configured as an external trigger to wake the software from its sleep mode.

An example of how a binary frame is configured in the look-up table is shown here:

```
CODE:   addwf PCL,F
1;
        retlw B'00100000'
        retlw B'00100000'
        retlw B'00000000'
        retlw B'00000000'
        retlw B'00100000'
        retlw B'00100000'
        retlw B'00000000'
        retlw B'00000000'
```

And here is a small extract of the assembly program listing:

```
START:  movlw 31        ;this is the total number of frames in the animation
        movwf FRAME     ;try to limit the number to 31

        clrf DISPL
        clrf MAP
        clrf COUNT

ST:     movf MAP,W
        movwf DISPL     ;display mapping starts at 0

ST1:    movf DISPL,W    ;move display variable into w
        call CODE       ;calls display byte from lookup table
```

```
              movwf PORTB         ;moves the lookup value to port B
              call PAUSE          ;pause to stabilize the display
              incf COUNT,W        ;increments count variable
              xorlw 8             ;does count = 8?
              btfsc STATUS,Z
              goto ST2            ;yes, a frame is complete
              incf COUNT,F        ;no, increment count
              incf DISPL,F        ;no, increment display variable
              goto ST1            ;go back to the start again

    ST2:      movf MAP,W          ;move map to w
              movwf DISPL         ;move w to display variable
              clrf COUNT          ;clear count
    KEY:      btfsc PORTB,0       ;is B0 being pressed?
              call ADJUST         ;yes, adjust speed
              decfsz REPEATF,F    ;decrease repeat, has it reached zero
              goto ST             ;no, keep showing the same frame

    NEXT:     movf FAST,W         ;yes, prepare to show the next frame
              movwf REPEATF       ;sets the repeat value to the FAST value again
              movlw 8
              addwf MAP,F         ;adds 8 to the map to shift to the next frame
              decfsz FRAME,F      ;have all animation frames been shown?
              goto ST             ;no, go and show the next 8-byte frame

              decfsz REPEAT,F     ;decrease animation repeat variable
              goto START
              goto SNOOZE         ;the animation has been shown 5 times go into sleep mode

    ADJUST:   movlw 2
              addwf FAST,W
              xorlw 40            ;does fast = max 40? = slow animation
              btfsc STATUS,Z
              goto RESET          ;yes, adjust speed to max
              movlw 2
              addwf FAST,F        ;no, increment speed
              return

    RESET:    movlw 2             ;yes, moves 2 to w
              movwf FAST          ;makes fast = 1 = very quick animation
              return
```

The Hex File

The file for the hex code that you need to program IC1 is called **LED Groovy Light Stick.hex**. The complete hex code listing is presented here:

```
:02000000FF28D7
:08000800FF28FF2882072034C5
:100010002034003400342034203400340034103D0
:10002000103400340034103410340034003408340F8
:100030008340034003408340834003400340434040
:1000400004340034003404340434003400340340340
:100050002034003400340234023400340342034DA
:100060002034083408340234203408340340343450
:10007000020340234023420342034023410234168
:1000800010340434043410341034043404340483488
:1000900008340834083408340834083408340434844
:1000A000434103410340434043410341034042362
:1000B000023420342034023402342034203420342034FA
:1000C00010340834043410340834103420342034318
:1000D000010340834043410340834103420342034308
:1000E0001034083404340834083410342034203480F8
:1000F0000303438343C343C343834303420343E348A
:100100002A342A343A3634A2A343E34363420434DD
:10011000103404340234023402340234103420340234F1
:10012000023408341034103408342340234002340340F9
:100130000003400340834083400340034003400340F
:100140000003413414341434103420003400340034AF
:10015003034223422342234223423400034003400434F8
:100160000034103414341434103400034003400348F
:10017000000340034083408340340034003420234ADD
:10018000143408341434234143408341434223422B
:1001900001434083414342234143408341434223411B
:1001A0000034143140034003084340034143400342A3455
:1001B000034143400342A3400341434003408445
:1001C0003E34003408343E34003408343E340834BD
:1001D0003E340834083408343E34083408343E3400A345
:1001E000234023408340834034003420342034D0D
:1001F0002A343E343634E3342A343E3436340730E2
:100200009F00831600308500013086000C0308100D9
:1002100008312143A020085018601130A09002903C
:10022000A4000530A5001F30A800A001A601A70169
:100230000026080A0002080820860042170A083A48
:100240000031925290A7AA0A0A1A29260A800A70130
:1002500006183521A40B18292908A40008300A60780
:10026000A80B1829A50B1329429029023029072A83A78
:1002700000319329D20230A90708000230A90080022F
:100280002208A1008B010B1D43290B11A10B43294F
:10029000080010308B00860163008B10053A05002C
:0202A000132920
:02400E00303F41
:00000001FF
```

Making It Move

After you have programmed IC1 and inserted it into the DIL socket, install the two AAA batteries into the battery holder. You should immediately see the row of five LEDs starting to flash and change position as the binary data in the look-up tables is presented to the LEDs. You can now attach the LED display lid to the enclosure using four screws. At this moment you're probably looking at the flashing LEDs and thinking, "What's so groovy about that?" Recall that the POV magic happens only after you start to move the groovy light stick—only then will you see the changing light patterns that the LEDs are creating. There are a few ways that you can make the light stick move, and you will find that the light patterns change depending on the method you use.

Shake It in the Air

Shaking the groovy light stick in the air in a dark room is the simplest and recommended method to see the light patterns. Simply hold the groovy light stick in your hand and wave it vigorously in front of you, and you should see an effect similar to that shown in Figure 19-11. Notice that altering the speed at which you shake the unit also alters how the light patterns look. Also try extending or reducing the size of the arc that you make when you wave the unit to see how the patterns change. You can also try pressing the push button on the side of the groovy light stick, as this will allow you to adjust the LED display speed. Once the LEDs have gone through their complete display sequence five times, IC1 goes into sleep mode and all five LEDs switch off. You can wake the unit up at any time by simply pressing the push button.

Figure 19-11 Shaking it

fixing a circular piece of plastic onto the back of the enclosure (in the center) and then rotating it on a smooth plastic surface. I drilled a hole into the plastic surface to prevent the enclosure from falling over when being rotated.

An alternative method is to attach the back of the enclosure to a small, battery-powered, variable-speed rotating motor. If you do this you will need to use suitable fixings screwed into the back of the enclosure so that the unit does not go flying into the air when it rotates. This method generates groovy light patterns that appear in a full circle and look really impressive, such as the example shown in Figure 19-12. Altering the motor speed also alters how the light patterns look.

Figure 19-12 Rotating the light stick using a motor

CAUTION If you decide to attach the groovy light stick to any rotating or moving object, first make sure that the enclosure is firmly attached to suitable brackets or fixings, and that these are also securely attached to the moving part. Also make sure that the plastic head (and the nut) of the switch, the four enclosure screws, the LEDs and their clips will not fly off when it is rotated. Rotating motors and bike wheels are dangerous, so approach the following methods with extreme caution. Do not attempt to touch the groovy light stick, or press the switch, when it is rotating. If you decide to install this project on your bicycle, please observe your local legal requirements relating to bike illumination, and ensure that this type of flashing LED display is acceptable for use on public streets and roads. Also note that this project is not intended for use as safety illumination for your bicycle.

Rotate It

The light patterns look really neat if the light stick is rotated 360 degrees at a fast enough speed. The simplest and safest method of doing this is to find a way of manually rotating the enclosure like a roulette wheel. I experimented with this method by

Bike Wheel

You might also consider attaching the groovy light stick to a wheel of your bike, making sure that the enclosure is firmly attached to multiple spokes of the wheel, so that it is unable to move or fly off. You will also need to ensure that it does not come into contact with any of the bike's parts (such as the forks or the brakes) when the wheel is rotating. This method produces a lighting effect like the image shown in Figure 19-13. Note that you may

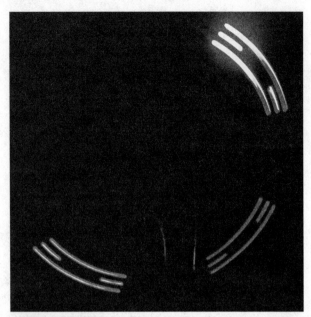

Figure 19-13 Rotating the light stick on a bike wheel

need to use a smaller, weatherproof enclosure if you decide that you want to make this a permanent fixture on your bike.

Glove LEDs

Another idea that I had, which I think is worth further investigation, is to mount the five LEDs onto the five separate fingers of a glove. The electronics can remain housed in the enclosure, and would be connected to the five LEDs via interconnecting wires; you could then hide the enclosure in your pocket or attach it to your forearm. The idea then would be to switch the unit on and wave your gloved hand vigorously to see the light patterns moving in the air. You would need to find a suitable method of fixing the LEDs securely to the glove, and you would probably also need to experiment a little to find the best position to mount the LEDs on the fingers (or fingertips) of the glove, to produce the best POV effect.

Further Modifications

If you know how to write your own assembly program or modify the assembly program provided here, you could design your own groovy light patterns by altering the binary code in the program's look-up table. A couple of other circuit modifications worth considering are to use seven LEDs rather than five, or to use tricolor LEDs instead of single-color versions. If you decide to experiment with five tricolor LEDs, you will need to utilize Port A and modify the circuit design accordingly.

CHAPTER 20

Showing Numbers on a Dot-Matrix Display: Dot-Matrix Counter

IF YOU HAVE ALREADY COMPLETED PROJECT 6 (mini digital display scoreboard) and project 14 (three-digit counter), you have built two types of digital counter, both of which use standard seven-segment LED displays to generate the numbers. Using seven-segment displays works fine in these applications, but one of the downsides is that the ways in which the numerical digits can be displayed are fairly limited. The project presented in this chapter shows you an alternative display method that uses a single, compact, 5 × 7 dot-matrix LED display like the one shown in Figure 20-1. The circuit uses persistence of vision (POV) to display the numerals, and you can select between two slightly different character sets via a push button.

Project 19
Dot-Matrix Counter

If you have read Chapter 17 and built the digital oscilloscope screen, you will have already seen how we can use dot-matrix LED displays to produce POV effects. This is the first dot-matrix LED display project in this book that allows us to generate readable numbers by using POV. For more details about POV concepts it is worth reading Chapter 15.

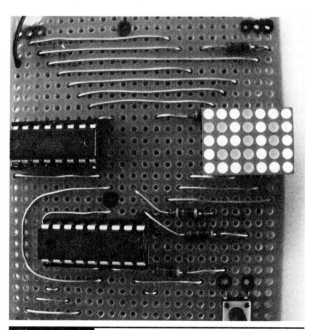

Figure 20-1 Dot-matrix counter project

PROJECT SPECIFICATIONS
■ A dual-digit counter counts from 0 to 99.
■ The dot-matrix display comprises 35 LEDs arranged in a 5 × 7 LED matrix (0.7 inch).
■ The display offers two different selectable character sets.
■ A single push button is used to increment the count, reset the counter, and select the character set.
■ The supply voltage is 3 volts.

How the Circuit Works

The circuit diagram for the dot-matrix counter is shown in Figure 20-2. At the heart of the diagram is a PIC16F628 microcontroller (IC1), which is connected to a small, 0.7-inch common anode dot-matrix display (5 × 7 LEDs).

The circuit diagram shows that the positive drive outputs of Port B (RB0–RB6) of IC1 are connected directly to pins 1 to 7 of the ULN2003, and these are converted to negative outputs on pins 10 to 16. These negative outputs are able to sink the current from each column of the common anode LED matrix. The positive outputs from Port A are fed through their individual resistors, R1–R5. The Port A configured outputs used in this design are RA0, RA1, RA2, RA3, and RA6. My original prototype used RA4 instead of RA6, but

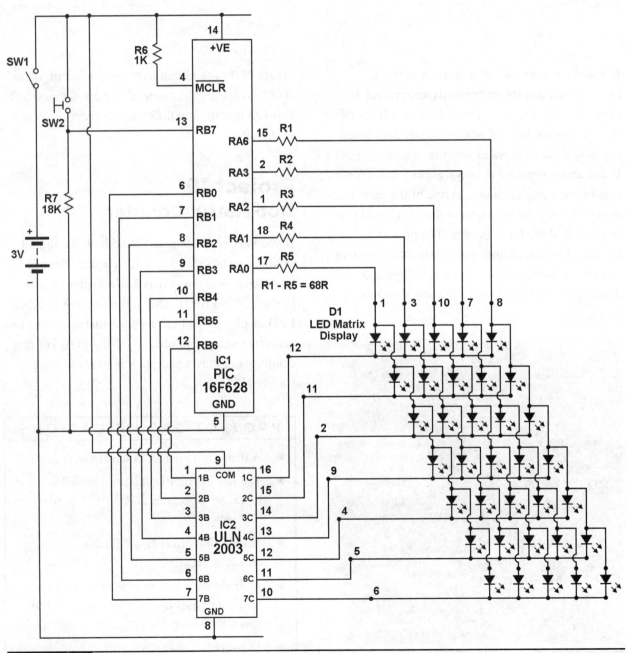

Figure 20-2. Circuit diagram for the dot-matrix display

output RA4 requires an additional pull-up resistor to provide an output. Because this design uses the microcontroller's internal 4-MHz crystal oscillator and does not require an external clock oscillator, output RA6 is available, so I decided to use this output instead. Ports RA4 and RA7 are not used and thus are configured as outputs in the software and are left unconnected.

The maximum current output allowed to be sourced from each of IC1's outputs is 25mA, so resistors R1 to R5 are set to limit the LED current to less than 20mA when using a 3V supply. The maximum current that can be sunk from each of IC1's outputs is also 25mA and, bearing in mind that a character column could require all five Port A outputs to be switched on, this could potentially draw over 100mA through a Port B output when it is activated. Therefore, a method of current buffering is required for each of the Port B outputs. I initially considered using seven individual transistors, but I wanted to make a neat stripboard layout, so I decided to use a ULN2003 buffer chip (which you may have encountered already if you completed the project in Chapter 9 and/or the project in Chapter 16). This chip is housed in a standard 16-pin IC DIL package and is able to switch up to 500mA, which is plenty for this application.

SW1 is the power switch. SW2 is a normally open (NO) push button switch, which is monitored by the software and is used both to increment the counter and to reset the count to 00. SW2 is connected to Port RB7, which is configured as an input and is normally held low via resistor R7; when the button is pressed, Port RB7 goes high.

Parts List

NOTE The Supplier and Part Number column of the following table lists specific parts that I used in this project. Refer to the appendix for additional details about acquiring your parts.

The parts you'll need for the dot-matrix counter project are listed in the following table.

PARTS LIST			
Code	Quantity	Description	Supplier and Part Number
IC1	1	PIC16F628-04/P Microcontroller	RS Components 379-2869 (Mfr: Microchip Technology Inc. PIC16F628-04/P)
IC2	1	ULN2003AN Darlington transistor array	RS Components 436-8451
R1-R5*	5	68Ω 0.5W ±5% tolerance carbon film resistor	—
R6	1	1KΩ 0.5W ±5% tolerance carbon film resistor	—
R7	1	18KΩ 0.5W ±5% tolerance carbon film resistor	—
D1	1	0.7-inch 5 × 7 common anode HE red dot-matrix LED display V_F (typical) = 2V, I_F (typical) = 20mA	RS Components 247-3141 (Mfr: Kingbright TA07-11EWA)
SW1	1	Single-pole panel mount toggle switch, 2A rated	RS Components 710-9674
SW2	1	0.24" × 0.24" (6mm × 6mm) tactile momentary push-to-make switch 0.67" high (17mm), 50mA rated	RS Components 479-1463 (pack of 20)

(continued)

226 Brilliant LED Projects

		PARTS LIST *(continued)*	
Code	Quantity	Description	Supplier and Part Number
Hardware	1	Stripboard, 0.1" (2.54mm) hole pitch, 37 holes wide by 24 tracks high	—
Hardware	1	18-pin DIL socket	—
Hardware	1	16-pin DIL socket	—
Hardware	1	20-way turned pin SIL socket (cut to size; see text)	RS Components 267-7400 (pack of 5)
Hardware	1	AAA battery holder (two AAA batteries)	RS Components 512-3552
Hardware	2	AAA battery (1.5V)	—

* Note: If you use a dot-matrix LED display that has different V_F and I_F values to those that are in the parts list, then you may need to alter the resistance and wattage values of these LED series resistors. Please refer to Chapter 2, which explains how to do this, and also consider the maximum output current capabilities of IC1, which are outlined in Chapter 12.

Stripboard Layout

The stripboard layout for the dot-matrix counter is shown in Figure 20-3.

The complete circuit is built on a single piece of stripboard that is 37 holes wide by 24 tracks high. Before you solder the components in place, you need to make 26 track cuts, including those

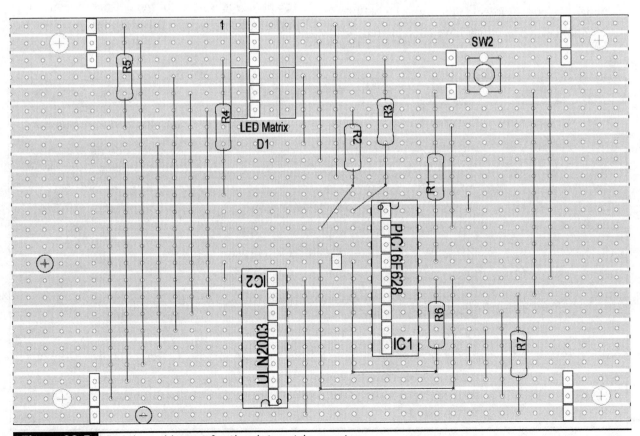

Figure 20-3 Stripboard layout for the dot-matrix counter

underneath the DIL sockets, plus a further 12 track cuts near the four mounting holes, which you need to drill in each corner. The track cuts are shown as white rectangular blocks on the stripboard layout.

How to Build and Test the Board

NOTE Please refer to Chapter 1 for soldering tips and techniques and for generic stripboard building guidelines.

Build the stripboard by carefully following the diagram shown in Figure 20-3. Note that the dot-matrix display (D1) is not soldered directly to the stripboard; instead, you need to solder two 6-way turned pin SIL sockets (cut from a 20-way SIL socket) into place and plug the display D1 into those sockets. You also need to make sure that the dot-matrix display is fitted in the correct orientation so that pin 1 matches the position on Figure 20-3. The underside of the dot-matrix display that I used is shown in Figure 20-4, and you will notice that there is a raised dimple on one side of the display which denotes the edge that contains pins 7 to 12. If you use a different dot-matrix display to the one in the parts list it may have different pin-outs to the one that I used, and if this is the case then you will need to modify the stripboard layout to suit your display. Always check the manufacturer's data sheet to confirm the pin configuration of your dot-matrix display.

The completed stripboard should look like Figure 20-5.

After you have built the stripboard, you can perform some checks to ensure that the correct voltages are present at the relevant pins of the DIL sockets for IC1 and IC2. You can also check the operation of each of the 35 LED dots on the display before inserting IC1 into place. To do this, you first need to fit IC2 into its DIL socket (noting its upside-down orientation on Figure 20-3). Then, connect a positive voltage to pin 17 of IC1's DIL socket and apply a positive voltage to pin 6 of IC1's DIL socket. This should illuminate the bottom-right LED on the circuit diagram. You can then keep power applied to IC1's DIL socket pin 17 and apply a positive voltage to IC1 socket pins 6 to 12 in turn. After you have tested the first row, repeat the process by applying a positive voltage to IC1 socket pin 18 and testing the next row of LEDs by again testing pins 6 to 12 in turn. You can continue this process by following the circuit diagram until you have proved that all 35 LEDs operate as expected.

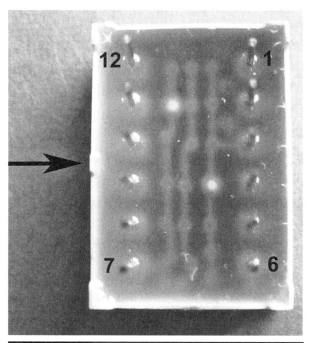

Figure 20-4 The pin configuration of the dot-matrix display that I used (the arrow shows the position of the raised dimple)

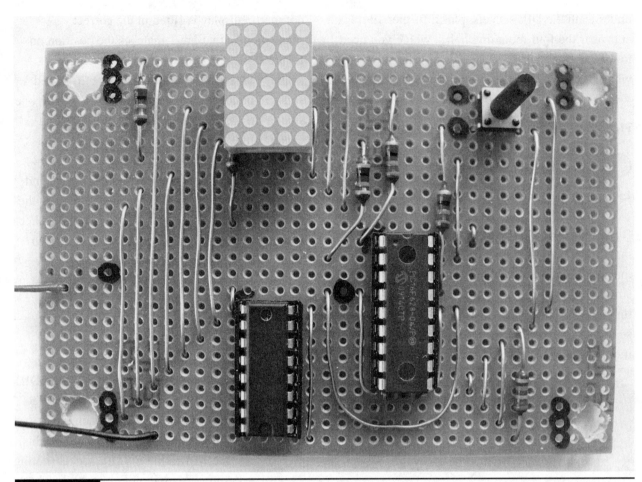

Figure 20-5 The completed dot-matrix counter stripboard

The PIC Microcontroller Program

Now you need to program IC1 and install it on the stripboard. You can download the assembly program and Hex file from the McGraw-Hill website, at www.mhprofessional.com/computingdownload. You can then program IC1 with the Hex file called **LED Dot Matrix Counter.hex**, as explained in Chapter 12.

The Assembly Program

The assembly program for this project is called **LED Dot Matrix Counter.asm**. The file includes detailed notes about how the program works, and the project in Chapter 21 provides a more detailed explanation of how the program makes the images appear on the dot-matrix display.

The Hex File

The file for the hex code that you need to download and program into IC1 is called **LED Dot Matrix Counter.hex**. The complete hex code listing is shown here:

```
:02000000632873
:080008006328628820700341D
:10001000043408340C341034143418341C342034B0
:100020002434820070E3411340E34003412341F3459
:1000300010340034193415341234003411341534AA
:100040000A340034073404341F34003413341534B4
:100050000D3400340E3415340934003401341D34A9
:100060000334003400A3415340A340034123415349D
:100070000E34003482071F3411341F34003400342E
:100080001F34003400341D341534173400341B457
:1000900015341F340034073404341F34003417344B
:1000A00015341D3400341F3415341D34003401342C
:1000B00001341F3400341F3415341F3400341B4073426
```

```
:1000C00005341F3400348601A001A201A101A4015E
:1000D000A501A901AA01AB01AC01AE0183168501FE
:1000E0008030860081308100831203303A3003230DB
:1000F0008400A1010430A8002908A7000F390620B8
:10010000A70027082E1811202E1C3A208000840AF0
:10011000A70AA80B81280430A8002A08A7000F39D5
:100120000620A70027082E1811202E1C3A20800038
:10013000840AA70AA80B9228A201A601A401A5017E
:100140000323084003230A400A101A628860101309B
:10015000A00024088400861BD020861FAB012C0839
:10016000FF3A0319E4202408A1002108073EA50056
:10017000A20A22082306031977288601000A87008F
:10018000271A27172708850020088600F820A00DC9
:10019000A10A210825060319A62821088400BD28E4
:1001A000AC0A2B180800AA0A2A080A3A0319DB2805
:1001B0002B14AC010800AA01A90A29080A3A03195C
:1001C000E4282B14AC010800A901AA012B142C0867
:1001D000FF3A0319ED20AC010800AC01861F0800AE
:1001E000AC0A01212C08FF3A031DEE28AE0A0800D4
:1001F0000230AD008B010B1DFB280B11AD0BFB2852
:10020000080000A30AD008B010B1D04290B11AD0B4A
:0402100004290800B5
:02400E00303F41
:00000001FF
```

Time to Count

Once you have programmed IC1 with the hex code, you can insert it into its DIL socket on the stripboard and apply the 3V supply. If you turn the stripboard on its side, you should immediately see two zeros (00) appear on the display, as shown in Figure 20-6.

If you now press and release SW2, the count will increment by one, to 01. This counting will continue each time you press SW2 until the count reaches 99 and then it will start at 00 again. If you press and hold SW2 for a few seconds, the counter goes blank; if you release SW2 at this point, the display will reset to 00.

Now for the interesting feature. If you continue to hold the button a few seconds longer after the display has gone blank, the character set for the numerals changes to the second character set. You

Figure 20-6 The display initially shows 00.

can then start to use the counter using the new character set. You can also toggle between the two character sets by holding down SW2 as just described. The two different numerical character sets that are programmed into IC1 are shown in Figure 20-7. On power-up, the dot-matrix counter defaults to character set #2.

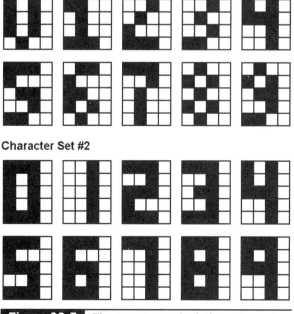

Figure 20-7 The two numerical character sets

> **NOTE** The microcontroller circuit layout for this project doesn't contain a decoupling capacitor across the + and − battery supply to smooth the supply voltage and to help avoid potential spurious triggering of the circuit. I didn't include a decoupling capacitor because the circuit worked well without one. If you experience a problem with your project, try fitting a 100nF or 0.1μF (minimum 10V rated) capacitor across the positive and ground rails of the circuit to see if that helps to alleviate the issue.

Enclosure and Alternative Uses

If you want to use the dot-matrix counter as a simple counter or scoreboard, you can mount the circuit into a suitably sized enclosure, which you can buy from an electronics supplier. You could also easily modify the stripboard layout to accept a digital input signal rather than a switch to increment the counter.

Chapter 21 shows you a totally different use for this circuit design and provides a more detailed explanation of how the circuit and display software operates.

CHAPTER 21

Creating Animations and Scrolling Text on a Dot-Matrix Display: Moving Message Destiny Predictor

THIS PROJECT TAKES ILLUMINATING A DOT-MATRIX display using POV to another level by showing you how you can generate moving messages and animations on a single dot-matrix display. The illusion of POV is explained in more detail in Chapter 15. The idea for this chapter's project was inspired by the fortune-telling machines found at many video arcades, amusement parks, and fairs, the type that you drop a coin into and ask a question, and from which you then receive an answer to discover your destiny. For example, you might ask, "Will my girlfriend marry me?" and the machine will then give a random message to deliver your fate, such as "No way!" The Moving Message Destiny Predictor, shown in Figure 21-1, does not promise to give you an accurate prediction of your future, but it does provide any one of a variety of short, generic answers when its button is pushed, making it a source of light-hearted fun and amusement at home and at parties.

This chapter also delves into the details of how the software generates the alphanumeric characters and makes them scroll across the display. It also explains how you can make a slightly unusual enclosure to house the electronics.

Figure 21-1 The Moving Message Destiny Predictor

Project 20
Moving Message Destiny Predictor

The project described here uses the same circuit design and stripboard layout as the previous project outlined in Chapter 20; this project, however, uses a totally different hex code that you need to program into the PIC Microcontroller. If you have not already built the dot-matrix counter in the last chapter, then don't worry because details of how to build the board are also explained in this chapter.

> **PROJECT SPECIFICATIONS**
>
> - The dot-matrix display comprises 35 LEDs arranged in a 5 × 7 LED matrix (0.7 inch).
> - A simplified 39 alphanumeric character set is programmed into the microcontroller.
> - There are 15 different 16-character predictions which can scroll across the display.
> - There is a short 28-frame animation sequence which is displayed while the prediction is being made.
> - The electronics are mounted in an unusual clear acrylic enclosure.
> - The supply voltage is 3 volts.

How the Circuit Works

The circuit diagram for this project is shown in Figure 21-2. If you previously completed the dot-matrix counter project presented in the previous chapter, you'll notice that this project has exactly the same circuit layout. At the heart of the diagram is a PIC16F628 microcontroller (IC1) that is connected to D1, which is a small, 0.7-inch common anode dot-matrix display (5 × 7 LEDs).

The circuit diagram shows that the positive drive outputs of Port B (RB0–RB6) of IC1 are connected directly to pins 1 to 7 of the ULN2003, and these are converted to negative outputs on pins 10 to 16. These negative outputs are able to sink the current from each column of the common anode LED matrix. The positive outputs from Port A are fed through their individual resistors, R1–R5. The Port A configured outputs used in this design are RA0, RA1, RA2, RA3, and RA6. My original prototype used RA4 instead of RA6, but output RA4 requires an additional pull-up resistor to provide an output. Because this design uses the microcontroller's internal 4-MHz crystal oscillator and does not require an external clock oscillator, output RA6 is available, so I decided to use this output instead. Ports RA4 and RA7 are not used and thus are configured as outputs in the software and are left unconnected. The maximum current output allowed to be sourced from each of IC1's outputs is 25mA, so resistors R1 to R5 are set to accommodate this when using a 3V supply.

NOTE Resistors R1 to R5 in this project (and the project in Chapter 20) are calculated to limit the LED current to a maximum of 20mA when using the dot-matrix display outlined in the parts list.

The maximum current that can be sunk from each of IC1's outputs is also 25mA and, bearing in mind that a character column could require all five Port A outputs to be switched on, this could potentially draw over 100mA through a Port B output when it is activated. For this reason I used a ULN2003 transistor array chip to switch each of the seven LED columns (you may have encountered the ULN2003 already if you completed the project in Chapter 9 and/or the project in Chapter 16).

Chapter 21 ■ Creating Animations and Scrolling Text on a Dot-Matrix Display

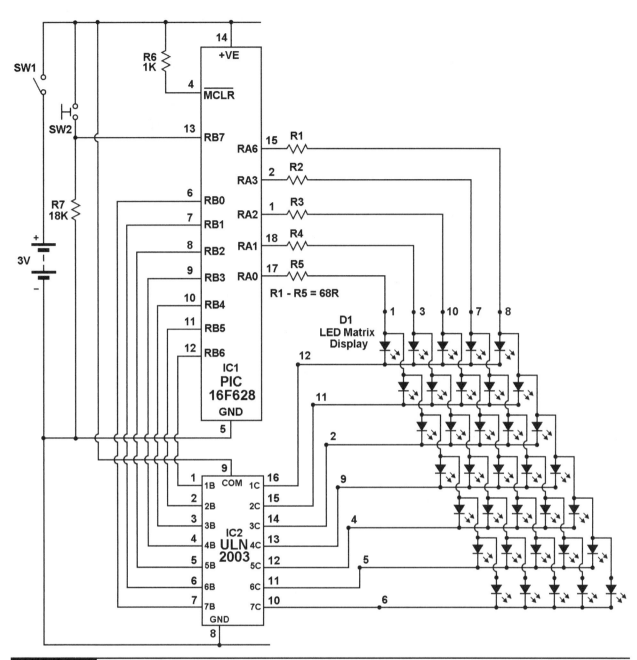

Figure 21-2 Circuit diagram for the Moving Message Destiny Predictor

SW1 is the power switch. SW2 is a normally open (NO) push button switch, which is monitored by the software and is used as the "Ask me a question" button. Pressing this button causes the microcontroller to choose a random message to show on the display. SW2 is connected to Port RB7, which is configured as an input and is normally held low via resistor R7; when the button is pressed, Port RB7 goes high.

Parts List

NOTE The Supplier and Part Number column of the following table lists specific parts that I used in this project. Refer to the appendix for additional details about acquiring your parts.

The parts you'll need for the Moving Message Destiny Predictor project are listed in the table that follows.

PARTS LIST

Code	Quantity	Description	Supplier and Part Number
IC1	1	PIC16F628-04/P Microcontroller	RS Components 379-2869 (Mfr: MicrochipTechnology Inc. PIC16F628-04/P)
IC2	1	ULN2003AN Darlington transistor array	RS Components 436-8451
R1–R5*	5	68Ω 0.5W ±5% tolerance carbon film resistor	—
R6	1	1KΩ 0.5W ±5% tolerance carbon film resistor	—
R7	1	18KΩ 0.5W ±5% tolerance carbon film resistor	—
D1	1	0.7-inch 5 × 7 common anode HE red dot-matrix LED display V_F (typical) = 2V, I_F (typical) = 20mA	RS Components 247-3141 (Mfr: Kingbright TA07-11EWA)
SW1	1	Single-pole panel mount toggle switch, 2A rated	RS Components 710-9674
SW2	1	0.24″ × 0.24″ (6mm × 6mm) tactile momentary push-to-make switch 0.67″ high (17mm), 50mA rated	RS Components 479-1463 (pack of 20)
Hardware	1	Stripboard, 0.1″ (2.54mm) hole pitch, 37 holes wide by 24 tracks high	—
Hardware	1	18-pin DIL socket	—
Hardware	1	16-pin DIL socket	—
Hardware	1	20-way turned pin SIL socket (cut to size; see text)	RS Components 267-7400 (pack of 5)
Hardware	1	AAA battery holder (two AAA batteries)	RS Components 512-3552
Hardware	2	AAA battery (1.5V)	—
Hardware	1	Enclosure (refer to the text in this chapter for more information)	—

* Note: If you use a dot-matrix LED display that has different V_F and I_F values to those that are in the parts list, then you may need to alter the resistance and wattage values of these LED series resistors. Please refer to Chapter 2, which explains how to do this, and also consider the maximum output current capabilities of IC1, which are outlined in Chapter 12.

Stripboard Layout

The stripboard layout for the Moving Message Destiny Predictor is also the same as in the previous project, as shown in Figure 21-3. Before you solder the components in place, you need to make 26 track cuts, including those underneath the DIL sockets, plus a further 12 track cuts near the four mounting holes, which you need to drill in each corner. The track cuts are shown as white rectangular blocks on the stripboard layout.

How to Build and Test the Board

NOTE Please refer to Chapter 1 for soldering tips and techniques and for generic stripboard building guidelines.

Build the stripboard by carefully following the diagram shown in Figure 21-3.

Note that the dot-matrix display (D1) is not soldered directly to the stripboard; instead, you need to solder two 6-way turned pin SIL sockets

Chapter 21 ■ Creating Animations and Scrolling Text on a Dot-Matrix Display 235

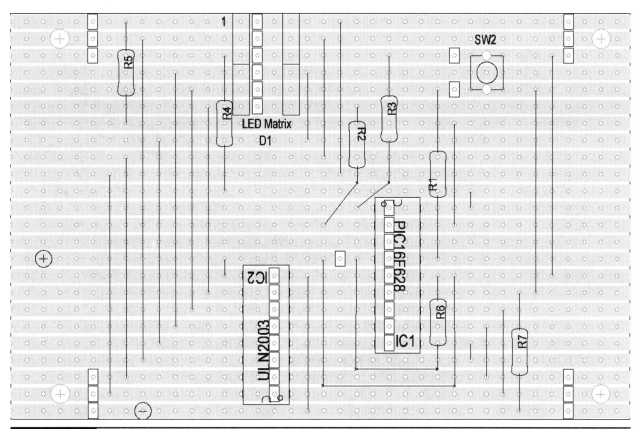

Figure 21-3 Stripboard layout for the Moving Message Destiny Predictor

(cut from a 20-way SIL socket) into place and plug the display D1 into those sockets. You also need to make sure that the dot-matrix display is fitted in the correct orientation so that pin 1 matches the position on Figure 21-3. The underside of the dot-matrix display that I used is shown in Chapter 20 (Figure 20-4), and you will notice that there is a raised dimple on one side of the display which denotes the edge that contains pins 7 to 12. This means that the raised dimple of the display should be facing towards the right-hand side of the board if you look at Figure 21-3. If you use a different dot-matrix display to the one in the parts list it may have different pin-outs to the one that I used, and if this is the case then you will need to modify the stripboard layout to suit your display. Always check the manufacturer's data sheet to confirm the pin configuration of your dot-matrix display.

The completed stripboard should look like Figure 21-4.

After you have built the stripboard, you can perform some checks to ensure that the correct voltages are present at the relevant pins of the DIL sockets for IC1 and IC2. You can also check the operation of each of the 35 LED dots on the display before inserting IC1 into place. To do this, you first need to fit IC2 into its DIL socket (noting its upside-down orientation on Figure 21-3). Then, connect a positive voltage to pin 17 of IC1's DIL socket and apply a positive voltage to pin 6 of IC1's DIL socket. This should illuminate the bottom-right LED on the circuit diagram. You can then keep power applied to IC1's DIL socket pin 17 and apply a positive voltage to IC1 socket pins 6 to 12 in turn. After you have tested the first row, repeat the process by applying a positive voltage to IC1 socket pin 18 and testing the next row of

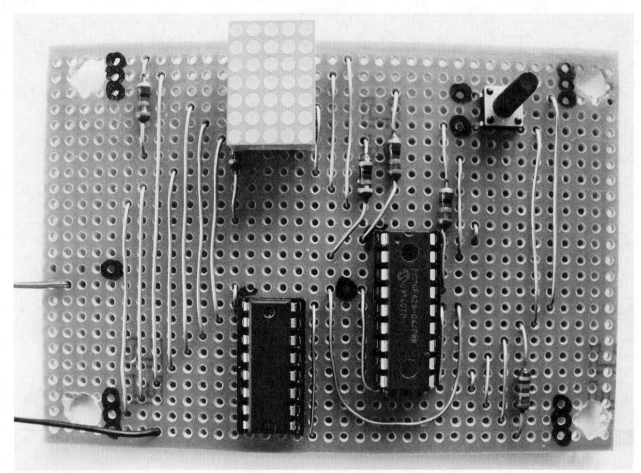

Figure 21-4 The completed Moving Message Destiny Predictor stripboard

LEDs by again testing pins 6 to 12 in turn. You can continue this process by following the circuit diagram until you have proved that all 35 LEDs operate as expected.

NOTE The microcontroller circuit layout for this project doesn't contain a decoupling capacitor across the + and − battery supply to smooth the supply voltage and to help avoid potential spurious triggering of the circuit. I didn't include a decoupling capacitor because the circuit worked well without one. If you experience a problem with your project, try fitting a 100nF or 0.1μF (minimum 10V rated) capacitor across the positive and ground rails of the circuit to see if that helps to alleviate the issue.

The PIC Microcontroller Program

Now you need to program IC1 and install it on the stripboard. You can download the assembly program and Hex file called **LED Destiny Predictor.hex.** from the McGraw-Hill website, at www.mhprofessional.com/computingdownload. You can then program IC1 with the Hex file as explained in Chapter 12.

The Assembly Program

The assembly program for this project is too long to be shown here; however, its operation is explained in more detail shortly. The file is called **LED Destiny Predictor.asm** and also includes detailed notes of how the program works.

The Hex File

The file for the hex code that you need to download and program into IC1 is called **LED Destiny Predictor.hex**. The complete hex code listing is presented here:

```
:02000000ED28E9
:08000800ED28ED288207003409
:10001000043408340C341034143418341C342034B0
:10002000243428342C343034343438343C344034A0
:10003000443448344C345034543458345C34603490
:10004000643468346C347034743478347C34803480
:10005000843488348C349034943498348207003457
:100060000034003400341E3409341E3400341F348C
:1000700015340A3400340E3411340A3400341F3479
:1000800011340E3400341F341534113400341F344D
:100090000534013400340E3411340D3400341F346F
:1000A0000434013400341134113400341F342B
:1000B0000F34013400341F3404341B3400341F3433
:1000C0001034103400341F3402341F3400341E3412
:1000D00001341E3400340E3411340E3400341F3415
:1000E000053406340034063409341634003415341F3421
:1000F00005341A3400341234153409340034013410
:100100001F34013400340F3410340F3400340734FA
:100110001834073400341F3408341F3400341B34BF
:10012000004341B340034173414340F3400341934BD
:1001300015341334003400343411340E3400341234B8
:100140001F341034003419341534123400341348F
:1001500015340A34003407340434341F34003413434A3
:1001600015340D3400340E34153409340034013440A
:100170001D34033400340A3415340A3400341234843
:1001800015340E34003401341534023400340034 94
:1001900003340034003482070034103420340307
:1001A0000403450346034703480349034A034B034EF
:1001B0000C034D034E034F034820761347 346B34AB
:1001C00020346134203471347534653473347743 4BC
:1001D00069346F346E342034203483316850180 30 C6
:1001E00086008030810083122830A3001430A400E0
:1001F0008601A001A201A101A901A501A601AB01EF
:100200008A01AD011030AC002E3084008001840AD8
:100210002908073A03190E29A90A06290430A9005A
:10022000AD1F472902308A0021080022A8008A0158
:100230002808203A0319392928083F3A03193C298C
:100240002808273A03193F29281F42292813A812F2
:100250002808 0620A80028082E208000AA0 840A60
:10026000A80AA90B2B29A10A21082C0603194B293E
:100270000E290030A80028292530A800282926307A
:10028000A80028292812A8121B30A8072829210 80D
:10029000DC20A8001829A9018001840A2908073A4E
:1002A00003195429A90A4C29A201A701A501A601F5
:1002B0002E3084002E30A500A1010130A0002508B9
:1002C00084002508A1002108073EA600A20A2208F2
:1002D0002306031981298601861BF12A0008A8003C
:1002E000281A281728088500200886008921A00DD3
:1002F000AA0AA10A2108260603195D2921088400FB
:100300006B29A70AA2012708483A0319F828A50A69
:100310005D290130AB008B010B1D8C290B11AB0B40
:100320008C2908008A01861F9729AA0A93292A087E
:100330000F39A80028080F3A0319A8012808CB2074
:0E034000A1002108103EAC00AD17A901042950
:10040000820 76D346F3472346534203474346834 48
:10041000 61346E3420346C3469346B3465346C343C
:100420007934449342734 6D3420347334 6134 793469
:1004300069346E34673420346E346F3420342034A1
:10044000203464346534663469346E346934743409
:1004500065346C3479342034793465347334203421
:10046000203469347434273473342034 61342034B4
:100470007334613466346534203462346534 7434E2
:10048000203469342034703472346534643469340F
:1004900063347434203461342034 6E346F34203447
:1004A000203470346C346534613473346534203 4F2
:1004B0006134733 46B3420346134673461346934AB
:1004C0006E3479346534733342034793465347334 5C
:1004D000203479346534733420347934 653473 34 9A
:1004E000203469342734 6D34203473 34 6134 7934E2
:1004F00069346E34673420347934653473342034 8D
:10050000203469347434203477346F346E3427 34B3
:100510007342034683 4613470347034 6534 6E342B
:100520002034743472347934203461346734613463
:1005300069346E3420346C34613474346534723 40C
:1005400020346934203470347234653464346934 4E
:10055000633474342034 61342034 7934653473 3432
:10056000203 46E346F3474342034613420346334 76
:100570006834613 46E346334653420342034 2034 7C
:1005800020346934203468346134 7634653420345E
:100590006E346F34203461346E347334773465 34A0
:1005A00072346934743420347734 6934 6C346C3484
:1005B000203468346134 7034703465346E3420 34DF
:1005C00020346934203 46E346534653464342034 26
:1005D00074346F342034 7434683469346E346B345A
:1005E00020348601A001A201A101A501A601AB0151
:1005F00004308A00A201A701A501A601FF2A01 304B
```

```
:10060000A0002508A1002108073EA600A20A220892
:10061000240603192028B8601861F922921081D24F8
:10062000A800281A2817280885002008860089219 4
:10063000A00DAA0AA10A210826060319FF2A0B2BDE
:10064000A70AA20127081C3A0319FA2A270800243E
:10065000A500FF2A0130AB008B010B1D2D2B0B11C8
:06066000AB0B2D2B08007E
:10080000820700340734 0E3415341C3423342A3460
:10081000313438343F3446344D3454345B346234EC
:10082000693470347734 7E3485348C3493349A341C
:10083000A134A834AF34B634BD3482070034003458
:100840000340340034003400340034003400340E34F6
:100850000A340E3400340034003 41F34113411349F
:1008600011341F3400341F34113411341134113455
:1008700011341F3400341F341134113411341F3437
:1008800000340034003 40E340A340E3400340034A2
:1008900000340034003 40434003400340034043 4B0
:1008A0000034003400340034003404340A340434 96
:1008B0000034003400340434 0A3411340A3404346B
:1008C00003404340A34113400341134 0A3404344A
:1008D000A34113400340034003415340A34153429
:1008E0000034003400340434 0A3411340A3404343B
:1008F00003404340A341134 003411340A3404341A
:100900000A341134003400340034113 40A34113400
:1009100000340034003400340034113400340034 26
:100920000034003400340034003400340034103417
:1009300000340134003400340034103408341134ED
:10094000023401340034103408 3405340A341434C9
:100950000234013408340434 0234 0434 08340434D6
:100960000234043402340534 0A341434 08340434 B0
:10097000023401340834113 402341034 08340134A0
:10098000003410340034013400341 034003400 34A6
:100990000340034003400340034 0340034003 4B7
:1009A000003400340034003400340034 00340034A7
:1009B00000340034003400340034003400340034 97
:0409C00000340034CB
:02400E00303F41
:00000001FF
```

How to Create Moving Characters

The program written for this project contains fifteen 16-character preprogrammed messages that are selected randomly at the press of the "Ask me a question" button. The program contains some useful data-manipulation techniques and converts standard tabulated ASCII text into a format that is displayed and scrolled along the LED matrix screen. The program also utilizes the PCLATH function to allow various text and image tables to be positioned throughout the program, rather than in just the first 256 bytes of the program as is normally the case. The random messages are displayed on a small LED dot-matrix display that is seven LEDs wide by five LEDs high. Normally LED message displays utilize a conventional ASCII character size of five LEDs wide by 7 LEDs high. Due to the space restrictions of using a single 5 × 7 dot matrix, this design uses characters that are only four LEDs wide by five LEDs high, allowing two characters of text to appear on the screen at the same time.

As shown in the circuit diagram presented earlier in the chapter in Figure 21-2, IC1 uses five Port A outputs to drive the five vertical LED rows, and uses seven Port B outputs to drive the seven horizontal LED columns; this is shown in Figure 21-5.

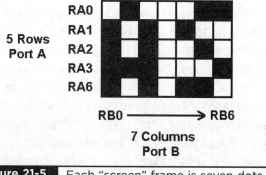

Figure 21-5 Each "screen" frame is seven dots wide by five dots high.

The program multiplexes the image on the dot-matrix "screen" by breaking down the stored messages and animations into individual columns, and this binary data is output to Port A in sequence. The seven Port B outputs are then switched on sequentially from left to right, which sinks the current from each of the Port A outputs. This means that only one column is switched on at

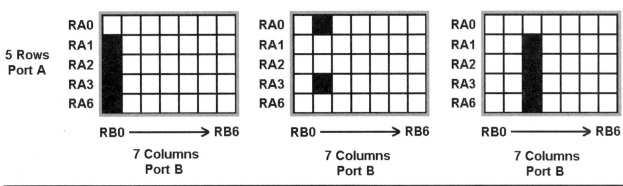

Figure 21-6 How each screen frame is built up

any one time, but the speed of this operation produces a POV effect and makes the image visible on the screen to the human eye. An example of how this is done for the first three columns of the start-up message ASK A QUESTION is shown in Figure 21-6.

The number of times that the same 7-byte-wide image is repeated on the screen is adjustable in the software by using a variable called RATE, which is set at 40 times for this project. Increasing this variable slows the rate of the moving image, and decreasing the variable increases the speed of the moving image.

How the Program Operates

I have tried to make the program as simple as possible so that you can easily manipulate it if you are familiar with assembly language programming. This means that you could alter the stored messages—or even the complete operation of the unit—if required. The assembly program includes plenty of notes that should help you to understand how it works. The program also uses the PCLATH function to allow text tables to be positioned throughout the program. There are three main areas of the program where these tables are contained:

- Start of the program (ORG 0)—Character set, position map, and start-up message

- Middle of the program (ORG 512)—fifteen 16-character messages

- End of the program (ORG 1024)—28-frame animation sequence and position map

The following is a basic outline of how the program operates:

1. On power-up, the unit shows a never-ending scrolling message saying ASK A QUESTION.

2. The player then asks a question out loud while pressing switch SW2.

3. Whenever switch SW2 is pressed, a random number variable, RANDOM, is selected. While the player keeps a finger pressed on SW2, a small animated sequence appears on the screen while the unit is "predicting" the answer.

4. When the player releases switch SW2, a string of direct addresses is populated with each character of the randomly selected 16-character message, and this message then scrolls along the screen.

5. The unit then shows the ASK A QUESTION message again.

6. The player can then ask another question by repeating the preceding operation sequence.

The Character Set

Before I explain how the messages and animations appear on the screen, we first need to look at the character set. This design uses a reduced character set of only 39 characters. The program converts each ASCII character in the message tables in the program into one of the characters shown in Figure 21-7. The characters are created in character tables in the program, starting with a space, which is character number 0, and then moving through the alphabet and numerals up to character number 38 for the apostrophe, shown in the lower-right corner of Figure 21-7.

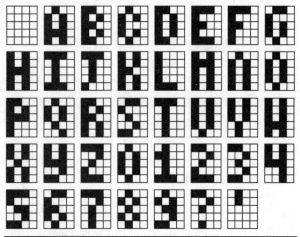

Figure 21-7 The character set

The program recognizes any text letter that is typed into the message tables of the assembly program and converts it into one of the characters in Figure 21-7. The program recognizes any of the following characters:

- Uppercase A–Z
- Lowercase a–z
- Numerical characters 0–9
- A space
- A question mark, ?
- An apostrophe, '

These characters are then converted into one of the 39 characters in the set shown in Figure 21-7. So, for example, typing either a lowercase letter a or an uppercase letter A into the message tables in the program will produce character A in the preceding character set (character number 1 in the character table). As mentioned previously, each character is only four dots wide by five dots high, and you will notice that the fourth column of each character has been left blank to create a space in between each character when the words of each message are generated.

Even though the characters are presented in a smaller format, four dots wide by five dots high, they have been designed so that they are easily readable as they scroll along the screen. As you probably noticed in Figure 21-7, the only two characters that are difficult to generate due to the reduced character width are the letters M and W. However, once you get used to the format, they are easily recognizable on the screen.

The format that the characters take in the assembly program tables is similar to that for the letters a to c shown in Figure 21-8. These are presented to Port A outputs of IC1 in turn, as explained shortly.

```
retlw %00011110 ;a
retlw %00001001 ;
retlw %00011110 ;
retlw %00000000 ;

retlw %00011111 ;b
retlw %00010101 ;
retlw %00001010 ;
retlw %00000000 ;

retlw %00001110 ;c
retlw %00010001 ;
retlw %00001010 ;
retlw %00000000 ;
```

Figure 21-8 How the letters a, b and c are created in binary code

If you rotate Figure 21-8 by 90 degrees counterclockwise, you will see how the number 1's in the binary code create each letter.

You will also notice that the preceding bytes of data do not activate Port RA6, but the program automatically copies the value of Port RA4 and copies it to bit 6. I did not use Port RA4 as an output because it requires a pull-up resistor when it is used as an output.

The 15 Prediction Messages

The fifteen 16-character prediction messages are stored in tables in ORG 512 and are listed here:

1. More than likely
2. Definitely yes
3. It's a safe bet
4. Yes Yes Yes Yes
5. I'm saying yes
6. I predict a yes
7. It will happen
8. I predict a no
9. It won't happen
10. Not a chance
11. I'm saying no
12. Please ask again
13. Try again later
14. I have no answer
15. I need to think

The 15 message by 16 character table fits nicely into the 256 blocks of program space that are allowed for table data. Increasing the quantity of messages or the length of each message could cause the program to crash or cause spurious faults, so be careful if you decide to play around with these parameters in this way.

While the program is running, the RANDOM variable is continuously incremented, and once the button is pressed, the value of RANDOM is selected and modified to limit it to a number between 0 and 14. This number is then used to select one of the 15 message predictions from the tables in the program, and the characters are populated into a string of direct addresses.

How Messages Are Made to Scroll on the Screen

The message is made to scroll along the matrix from right to left by the program shifting a 7-byte-wide window along the message, which is stored in the string of direct addresses. The direct addresses are populated once the random message is chosen, and there are spaces added to the start and end of the message so that the message scrolls seamlessly along the screen. This is demonstrated in Figures 21-9 to 21-11.

Frame 1 appears on the screen 40 times before shifting to frame 2, which appears on the screen 40 times before shifting to frame 3, which appears on the screen 40 times, and so forth. This continues until all frames in the direct addresses have been read, and then the sequence starts again. So, as you can see, the message is made to look as if it is moving along the screen but it is actually the 7-byte-wide window moving from left to right along the message that creates this visual effect.

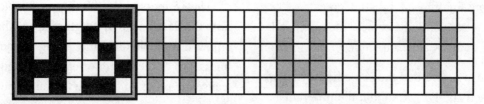

Figure 21-9 Frame 1 appears on the screen 40 times before shifting to...

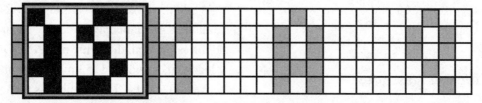

Figure 21-10 Frame 2 for 40 times and then to...

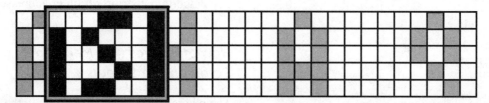

Figure 21-11 Frame 3 for 40 times... etc.

How the Animated Sequence Is Created

The short animated sequence, which is generated when SW2 is pressed, is stored in a separate table at the end of the program in ORG 1024 and is twenty-eight 7-byte frames long. The way that the animation is shown on the matrix is similar to that described for the moving message, except that the frame jumps 7 bytes at a time using a message map in the assembly program. There is also a separate variable called XRATE that controls the number of times that the image is repeated on the screen before it moves on to the next frame. XRATE is currently set to 20 in the program. The operation of the frame sequence is shown in Figures 21-12 to 21-14. Frame 1 is presented to the screen 20 times before moving to frame 2, which is a jump of 7 bytes along the animation table, and this process repeats for all 28 frames and continues to repeat for as long as SW2 is pressed.

The animation sequence that appears when the button is pressed does not use direct addresses; rather, the image is taken directly from the tables and presented to Port A directly.

Construction Details

As mentioned earlier in this chapter, the stripboard construction for this project is exactly the same as outlined for the dot-matrix counter project presented in Chapter 20. This section explains how I incorporated the electronics into an unusual enclosure. I decided to minimize costs, and to make the project look a little different, by using two pieces of 0.078" (2mm) thick clear acrylic to make a stripboard sandwich. Of course you may decide to mount the electronics in a standard enclosure if you so wish.

Before you build the enclosure and fit the electronics, you should program and fit IC1 to the

Chapter 21 ■ Creating Animations and Scrolling Text on a Dot-Matrix Display 243

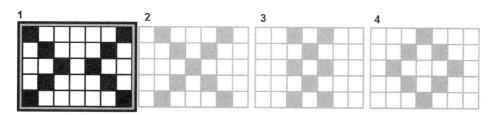

Figure 21-12 Frame 1 is presented to the screen 20 times before moving to...

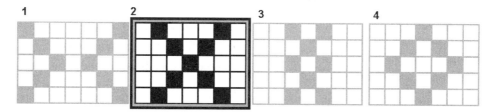

Figure 21-13 Frame 2, which is a jump of 7 bytes along the animation table...

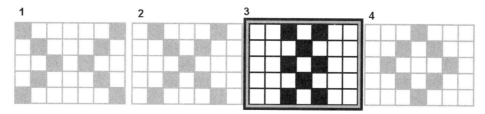

Figure 21-14 And this process repeats for all 28 frames and continues to repeat for as long as SW2 is pressed.

board and test that the circuit works okay. Powering the circuit up with a 3V supply should immediately activate the scrolling text message "ASK A QUESTION" on the LED matrix display. If this does not happen, then disconnect the battery immediately and perform the usual checks before proceeding any further.

If you decide to build an enclosure similar to the one I built, then you will need the following additional hardware:

- Two pieces of 0.078" (2mm) thick clear acrylic, 2.75" (70mm) wide by 5.3" (135mm) long—I purchased a sheet of clear acrylic from a hardware store

- 4 M3 acrylic screws 0.79" (20mm) long—similar to RS Components part number 527-656

- 16 M3 acrylic nuts—similar to RS Components part number 525-701

- One sheet A4 glossy photographic paper

- Cable ties, cable tie base, and double-sided adhesive tape

CAUTION Take extra care when drilling or cutting into acrylic because it can shatter and split easily. You can fit a 0.125" (3mm) drill bit into a wooden handle and make pilot holes into 0.078" (2mm) acrylic by turning the drill bit by hand. You can then use larger, sharp drill bits on a slow drill speed to get the desired hole size. Always wear safety glasses and take your time.

The Display Card

A nice feature of using a clear acrylic "sandwich" is that you can create a professional-looking front display by making a printed display card. I used a piece of printed photographic paper with a design that I created in Microsoft Publisher. You can download this from the McGraw-Hill website at www.mhprofessional.com/computingdownload; the file is called Destiny Predictor Display Card.

> **TIP** If you decide to make your own front display card, I recommend that you download my version and make sure that the holes on the display card that you design match up with those on my display card, so that the switches and dot-matrix display on the stripboard align properly with the holes on your card.

After you download and print the card, carefully cut a rectangular hole in the paper for the dot-matrix display by using a scalpel or box cutter and a metal ruler. You can cut the two holes for the switches into the card by using (ideally) a suitably sized hole punch. After you cut these holes, match up the display card to the stripboard layout and the pieces of clear acrylic and mark out on the acrylic the four hole positions for the stripboard and two further holes for switches SW1 and SW2. You can then drill the two pieces of acrylic accordingly. The front piece of acrylic will contain six holes, including two for the switches, and the rear piece of acrylic will contain four holes.

Next, using four M3 nylon screws and nuts and power switch SW1, you can mount the display card to the front piece of acrylic. Then, affix the AA battery holder to the rear piece of acrylic by using double-sided adhesive tape. The enclosure should now look like the example shown in Figure 21-15.

> **TIP** Mount the stripboard toward the top of the display card so that there is enough room for the battery holder to sit neatly between the stripboard and the power switch. I suggest that you make the fixing holes in the stripboard and acrylic pieces slightly larger than the M3 screws so that you have some wiggle room to adjust the position of the stripboard accordingly.

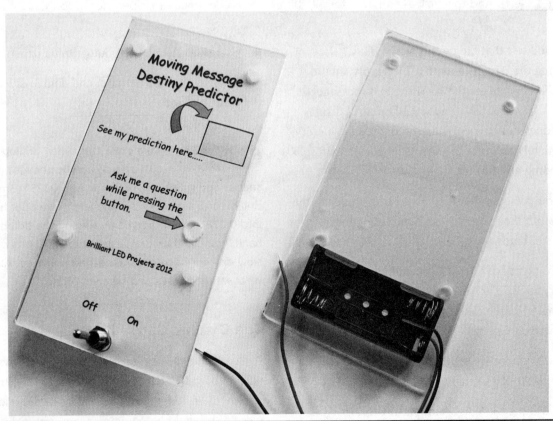

Figure 21-15 Preparing the two pieces of acrylic

Now fit another M3 nut partway down each M3 screw to provide rest points for the stripboard. You can then fit the stripboard to the screws and adjust the rest point nuts so that the matrix display lays on the acrylic and is visible through the rectangular hole in the display card. The long shaft of the push button should also protrude neatly through the hole on the front panel.

You can now fit a further M3 nut behind the stripboard to hold it into place, and then cut and solder the battery cables to the relevant points on the power switch and the stripboard. Insert the two AA batteries into the holder and make sure that the unit operates when you flick the power switch on.

Finally, fit the rear piece of acrylic, with the battery holder attached, to the M3 screws and secure the acrylic using the last four M3 nuts. The lower-right side of the final stripboard sandwich should look like the example shown in Figure 21-16.

The battery holder fits snugly between the power switch and the stripboard, so a bit of adjustment may be required to make sure that everything fits together neatly. Once you have finished constructing the unit, you should end up with a handheld device that looks like the photographs in Figures 21-17, 21-18, and 21-19.

Discover Your Destiny!

After you have built the Moving Message Destiny Predictor, you can turn it on using the power switch, which is mounted on the front face of the unit. In normal operation, the dot-matrix display shows a scrolling start-up message that says ASK A QUESTION. The player then asks a question out loud while pressing push button SW2. For the duration of the button press, an animated sequence plays continuously on the LED matrix while the device prepares to make a prediction. Releasing the button then shows one of the 15 random predictions scrolling along the display. The display then reverts to the ASK A QUESTION message and the player can ask another question about their future and repeat the preceding operation.

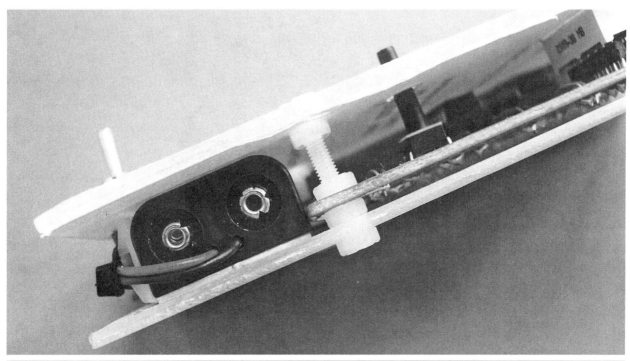

Figure 21-16 The display card and stripboard are sandwiched between the two pieces of acrylic like this.

Figure 21-17 A rear view of the Moving Message Destiny Predictor before the rear panel is fitted

Figure 21-18 A rear view with the panel fitted

Figure 21-19 A front view of the Moving Message Destiny Predictor

Possible Program Modifications

As previously mentioned, I have tried to make the circuit design and the program as simple as possible so that you can easily manipulate them to alter the stored messages or even the complete operation of the unit. Along those lines, this section provides some other possible project ideas. Although I have not tried out these modifications, modifying the operation to accommodate these changes should be fairly straightforward for anyone who is familiar with PIC Microcontroller programming.

Here are some ideas:

- An inexpensive mini-message display that could be made to display all 15 preprogrammed messages (of your choosing) one after the other in a continuous loop.

- A scrolling digital clock that shows the time as it moves across the screen. The circuit may need to be modified to include an external crystal oscillator to create accurate clock timings.

- A mini-matrix display that can show moving animated images.

A Final Word from Me

Our journey to discover some of the brilliant things that we can do with LEDs has come to an end for now. I really hope that you have enjoyed building and playing around with the projects in this book as much as I enjoyed designing and building them. I would be really interested in hearing from readers who have been inspired by the projects in this book, and you are welcome to share your ideas through my Facebook page, Brilliant LED Projects. Until the next time, I wish you well in your adventures in electronics!

APPENDIX

Useful Resources

THIS APPENDIX OUTLINES SOME OF THE RESOURCES that I have found useful as an electronics hobbyist and also when writing this book.

Electronic Components Suppliers

The parts listed for each of the projects in this book can be purchased from reputable electronic component suppliers. There are plenty of electronic suppliers out there, both on the Web and locally in bricks-and-mortar stores, and you may have your own favorites that you use regularly. You'll notice in the Parts List table of each project that I've provided a specific supplier's name and part number for some of the parts. Those references are to the parts that I used in the project. Because I am located in the United Kingdom, I acquired my parts from United Kingdom suppliers, several of which are listed in this section (along with one United States counterpart). To avoid high shipping costs, you'll likely want to get your parts from a supplier located within your country. I've listed the parts I used simply as a reference to help you to identify and source specific parts. For each part I've listed, you can search by part number on the respective supplier's website to get a visual and further technical details.

It is also worth mentioning that the specifications of the various components outlined in the parts lists may vary among different manufacturers and ranges, and could be subject to change. Therefore you should always check the up-to-date manufacturer's datasheet of the component that you use; this is to ensure that the pin-outs and electrical specifications of your components match the requirements of the project that you are building.

RS Components Ltd
Various electronic and hardware components that are used in each of the projects, including many of the LEDs and ICs, were sourced from this supplier.
www.rswww.com
www.uk.rs-online.com

Allied Electronics, Inc.
United States distributor owned by same group that owns RS Components (note that RS Components part numbers are not the same as Allied Electronics part numbers).
www.alliedelec.com

ESR Electronic Components Ltd
Various sizes of stripboard are available from this supplier, including the two main sizes that are used in this book.
www.esr.co.uk

Maplin Electronics

The narrow enclosure that is used in three of the projects (the LED Flashlight, Light Sword, and Groovy Light Stick), and the 3 × AA (4.5 volt) battery holders were sourced from this supplier. www.maplin.co.uk

PIC Microcontroller Reference Books

McGraw-Hill Professional has an excellent range of technical books that cover many subjects, including how to program PIC Microcontrollers. You can find more information by going to www.mhprofessional.com and entering **PIC Microcontroller** in the Search box.

Electronics Hobbyist Magazines

There are quite a few hobbyist magazines devoted to electronics, and they are a great source of information and inspiration for anyone who is into electronics. *EPE Magazine* is the one magazine that I have read since I started building electronic circuits, and it has helped me to gain a great understanding of electronics over the years. I have also had projects and articles published in this magazine in the past. It is available in both online and print editions.

Everyday Practical Electronics (EPE) Magazine
www.epemag3.com

PIC Microcontrollers

These are really flexible microcontrollers, and this is the microcontroller brand that I used in all of the projects that required a microcontroller. They are manufactured by:

Microchip Technology Inc.

Datasheets and lots of useful information and resources are available on the company's website. www.microchip.com

LochMaster 4.0 Stripboard Software

I used the LochMaster 4.0 software package from ABACOM to design each of the stripboard layouts for the projects presented in this book. ABACOM kindly gave me their permission to reproduce the screenshots of the stripboard layouts that I designed; these are printed in each of the chapters in this book.

The software is available from: www.abacom-online.de/uk.

Brilliant LED Projects

Downloads for the relevant projects in this book can be found at www.mhprofessional.com/computingdownload.

Be sure to visit my Facebook page for this book, Brilliant LED Projects, and share your LED project experiences.

Index

− (cathode). *See* cathode (−)
. (decimal point), 164
+ (anode). *See* anode (+)
7-segment display LEDs. *See* seven-segment display LEDs
14-stage binary ripple counter, 112
15 prediction messages, 241
74HC range of ICs
   4060 binary ripple counter, 112
   experimental LED sequencer circuit, 85–86
   LED candle, 120
555 timer
   in bike flasher, 51
   clock generator, 42–44
   digital oscilloscope screen, 189
   digital oscilloscope screen project, 197–200
   in experimental LED sequencer circuit, 87
4017 counter
   in color-changing disco lights circuit, 98
   defined, 85–86
   experimental LED sequencer circuit, 86–87. *See also* experimental LED sequencer circuit
4026B IC, 73–74, 79
4060 counter
   in color-changing light box, 64–65
   experimenting with LED candle, 125
   LED binary ripple counter, 112
   LED candle, 120. *See also* flickering LED candle
4511B IC
   invisible secret code display, 152, 160
   seven-segment display LEDs, 73
7555 timer, 42, 59

A
A, 22
ABACOM LochMaster 4.0, 2, 250
acrylic drilling, 105, 243
afterimage, 163
Allied Electronics, Inc., 249
amber color, 182
amplifier, 199–200
amps calculations, 24. *See also* milliamps (mA)
angle of soldering iron, 10
animations, 242. *See also* Moving Message Destiny Predictor
anode (+)
   alternative circuit for bike flasher, 60
   backpack illuminator circuit, 174
   in color-changing disco lights, 104
   color-changing disco lights circuit, 98
   defined, 22
   groovy light stick, 214
   LED flashlight, 28
   RGB LEDs, 64
   soldering LED string, 145
   tricolor LEDs, 62
antistatic mats, 12
antistatic precautions, 12
ASCII characters, 240–241
assembly code
   backpack illuminator, 181–183
   digital oscilloscope screen, 195–196
   dot-matrix counter, 228
   downloading, 14
   groovy light stick, 218–219
   LED light sword, 148
   low-res shadow camera, 207
   Moving Message Destiny Predictor, 236
   for PIC Microcontroller, 134–135
   three-digit counter, 170–171
astable circuits, 42

B
backpack illuminator
   circuit diagram, 175
   completing enclosure, 180
   how circuit works, 174
   how to build display board, 178–180
   how to build driver board, 177–178
   incorporating display into fabric, 184–186
   overview, 173
   parts, 176
   PIC Microcontroller program, 181–183
   stripboard layout, 177
   testing, 184
   testing boards, 180–181
batteries
   charging capacitors, 33, 39
   fault-finding stripboard circuit, 7
   flickering LED candle, 123
   "green" LED flashlight, 31
   holder for LED candle, 124
   illuminating multiple LEDs, 25
   LED candle, 120
   in LED flasher project, 47
   LED flashlight, 26
   LED light sword, 146–147
   mini digital display scoreboard, 75
   in rear LED flasher circuit, 53
   testing RGB LEDs, 68
   three-digit counter, 166
BC109 NPN transistors, 172
bicolor LEDs, 61–62
bike flasher. *See* LED bike flasher
bike wheel, groovy light stick on, 221–222
binary code for ASCII characters, 240–241
binary coded decimal (BCD)
   invisible secret code display, 152
   seven-segment display LEDs, 73
binary counting
   LED binary ripple counter, 116, 118
   three-digit counter display codes, 166–167
binary frames, 218
binary ripple counter. *See* LED binary ripple counter
block diagrams, 188
blue resistor color code, 15
breadboard
   connecting programmer to, 136
   using capacitor to power LED, 32–34
   working with, 13
brown resistor color code, 15

C
camera, experimental low-res shadow. *See* experimental low-res shadow camera
camera sensor mask, 208–210
cameras, digital, 151, 158–159

candle project. *See* flickering LED candle
capacitors
　555 timing calculations, 43
　bike flasher timing, 52
　in color-changing light box, 64
　decoupling, 98
　defined, 31–34
　experimenting with LED candle, 124
　experimenting with values, 49
　LED flasher project, 46–48
　mini digital display scoreboard, 75
cascade circuits, 85
cathode (−)
　backpack illuminator circuit, 174
　in color-changing disco lights, 104
　color-changing disco lights circuit, 98
　defined, 22
　groovy light stick, 214
　LED flashlight, 29
　seven-segment display LEDs, 72
　soldering LED string, 145
　tricolor LEDs, 62
characters
　dot-matrix counter, 229
　how to create moving, 238–239
　Moving Message Destiny Predictor, 240–241
　seven-segment display LEDs, 71–72
charge time (T1)
　555 timing formulas, 42–43
　rear LED flasher circuit, 52
charging capacitors, 32, 39
chase circuits, 85
chessboard pattern, 182
circuit diagrams
　555 timer, 42
　backpack illuminator, 175
　basic LED, 23
　bike flasher, 52
　clock generator, 41
　color-changing disco lights, 99
　color-changing light box, 64
　digital oscilloscope screen, 190
　dot-matrix counter, 224
　experimental LED sequencer circuit, 88
　"green" LED flashlight, 35
　groovy light stick, 212
　illuminating multiple LEDs, 25
　invisible secret code display, 153
　LED binary ripple counter, 113
　LED candle, 120
　LED connections, 22
　LED flasher, 45
　LED flashlight, 26
　LED light sword, 140
　LED scanner, 129
　low-res shadow camera, 202
　mini digital display scoreboard, 74

Moving Message Destiny Predictor, 233
　stripboard layouts and, 15–16
　three-digit counter, 165
　tips, 39
circuit symbols. *See* symbols
circuits
　backpack illuminator, 174
　breadboard capacitor, 32–34
　clock generator. *See* clock generator
　color-changing disco lights, 98
　color-changing light box, 63–65
　digital oscilloscope screen, 188–191
　dot-matrix counter, 224–225
　groovy light stick, 211–212
　how to build on stripboard, 2–8
　invisible secret code display, 152–153
　LED binary ripple counter, 112–113
　LED candle, 119–121
　LED light sword, 139–141
　LED scanner, 128–130
　low-res shadow camera, 202–203
　mini digital display scoreboard, 73–75
　Moving Message Destiny Predictor, 232–233
　possible modifications to color-changing light box, 70
　rear LED flasher, 52–53
　three-digit counter, 164–166
clash button, 141
clock, 247
clock circuits, 87
clock generator
　555 timer, 42–44
　basic single-LED flasher, 45–49
　overview, 41
CMOS ICs
　74HC range of ICs, 85–86
　4511B BCD, 152
　antistatic precautions, 12
　binary ripple counter, 112
　decoupling capacitors, 98
　experimenting with LED candle, 125
　seven-segment display LEDs, 73
code, assembly. *See* assembly code
code, hex. *See* hex code
code, invisible secret display. *See* invisible secret code display
color-changing disco lights
　circuit diagram, 99
　enclosures, 101
　how circuit works, 98
　how to build and test, 102–104
　how to build LED display, 104–109
　overview, 97
　parts, 98, 100–101
　stripboard layout, 101–102
　time to disco, 110
color-changing light box, 63–70

color codes
　display board, 179
　resistors, 15
colored LEDs. *See* multicolor LEDs
columns of LEDs
　digital oscilloscope screen, 188–189
　groovy light stick, 218
　Moving Message Destiny Predictor, 232, 238–239
components
　astable timings, 48–49
　breadboard circuits, 32
　fitting to stripboard, 5–6
　LED, 21
　soldering leads, 9–12
connections
　anode (+). *See* anode (+)
　cathode (−). *See* cathode (−)
　color-changing disco lights, 98
　LED illumination, 22
　seven-segment display LEDs, 95
constructing Moving Message Destiny Predictor, 242–243
consumption, current. *See* current consumption
copper wire, 145–146
counters
　4017 counter. *See* 4017 counter
　4026B IC, 73
　4060 binary ripple counter. *See* LED binary ripple counter
　dot-matrix, 223–230
　three digit, 164–172
crystal, 112–113
current consumption
　74HC range of ICs, 85–86
　4026B IC, 73
　4060, 112
　experimenting to reduce current consumption, 59–60
　low-res shadow camera, 203
　RGB LEDs, 64
　three-digit counter, 164, 166
current, sink. *See* sink current
current, source
　bike flasher, 51
　defined, 23, 44
cutting stripboard, 1, 3
cutting wire links, 5

D

datasheets, 24
daylight, 203
decimal point (DP), 164
decoupling capacitors
　defined, 98
　in groovy light stick, 212
　microcontroller and, 138, 149
desoldering tool
　defined, 9
　use of, 11–12

destiny predictor. *See* Moving Message Destiny Predictor
diagrams, circuit. *See* circuit diagrams
diffuser sleeve, 148
digital cameras, 151, 158–159
digital display scoreboard. *See* mini digital display scoreboard
digital oscilloscope screen
 how circuit works, 188–191
 how to build and test, 192–195
 other ideas, 199–200
 overview, 187–188
 parts, 191
 PIC Microcontroller program, 195–196
 stripboard layouts, 192
 waveforms, 196–199
DIL (dual in-line) sockets. *See* dual in-line (DIL) sockets
diodes
 in color-changing light box, 65
 defined, 21
 LED illumination and, 22
discharge time (T2)
 555 timing formulas, 42–43
 rear LED flasher circuit, 52
disco lights. *See* color-changing disco lights
display board
 backpack illuminator, 177, 178–180
 groovy light stick, 214–215
 incorporating into fabric, 184–186
 testing, 180–181
display card for Moving Message Destiny Predictor, 243–245
display codes for three-digit counter, 166–167
display scoreboard. *See* mini digital display scoreboard
dot-matrix counter, 223–230
dot-matrix display
 animations, 231
 counter, 223–230
 digital oscilloscope screen. *See* digital oscilloscope screen
 Moving Message Destiny Predictor. *See* Moving Message Destiny Predictor
downloading hex files, 134
DP (decimal point), 164
drilling
 acrylic, 243
 preparing secret code display, 155–156
 safety tips, 105
driver board
 backpack illuminator, 177–178
 testing, 180–181
dual in-line (DIL) sockets
 in color-changing light box, 68
 defined, 4

dot-matrix counter
 configuration, 227
 experimental LED sequencer circuit, 88, 90
 testing LED binary ripple counter, 116
dynamo, 59

E
electrolytic capacitors. *See also* capacitors
 defined, 31–34
 in flickering LED candle, 121
 in LED scanner, 129
electromagnetic compatibility (EMC), 16
electronic components, suppliers, 249–250
electronics
 miniaturization, 210
 warnings, 16–17
electronics hobbyist magazines, 250
EMC (electromagnetic compatibility), 16
enclosures
 backpack illuminator, 180
 color-changing disco lights, 101
 color-changing light box, 68–70
 dot-matrix counter, 230
 "green" LED flashlight, 39–40
 groovy light stick, 215–217, 221
 invisible secret code display, 155–156
 LED bike flasher, 57–59
 LED candle, 124
 LED flashlight, 27–30
 LED light sword, 142–144
 mounting low-res shadow camera, 208–210
 mounting mini digital display scoreboard, 80–82
 Moving Message Destiny Predictor, 242–243
environmental safety tips, 3
equipment for soldering, 8–9
ESR Electronic Components Ltd, 249
Everyday Practical Electronics (EPE) Magazine, 250
experimental LED sequencer circuit
 further modifications, 95–96
 how circuit works, 87–88
 how to build and test, 90–92
 overview, 86–87
 parts, 88–89
 stripboard layout, 88, 90
 time to experiment, 92–95
experimental low-res shadow camera
 further ideas, 210
 how circuit works, 202–203
 how to build and test, 204–206
 mounting board, 208–210

 overview, 201–202
 parts, 203–204
 PIC Microcontroller program, 207–208
 seeing shadow, 210
 stripboard layout, 204–205

F
F (frequency of oscillation)
 555 timing formulas, 43
 rear LED flasher circuit, 53
fabric backpack illuminator, 184–186
farads (F)
 defined, 32
 supercapacitors, 32
fault-finding
 color-changing disco lights, 109
 digital oscilloscope screen, 187
 "green" LED flashlight, 38
 LED binary ripple counter, 118
 stripboard circuit, 7–8
 tip, 39
fifteen prediction messages, 241
fire hazards
 nonflammable fabric for display, 185
 nonflammable insulation tape, 5
 soldering iron, 3
firmware, 128
flashers, 45–49. *See also* LED bike flasher
flashlight project
 basic LED, 25–30
 "green" LED flashlight, 34–40
flickering LED candle
 alternative IC, 125
 enclosure, 124
 experimenting with, 124–125
 how circuit works, 119–121
 how to build and test, 122–123
 overview, 119
 parts, 121
 stripboard layout, 122
flux, 9
formulas
 555 astable timing, 42–43
 binary ripple counter frequency, 112
 digital oscilloscope screen, 189–190
 series resistor calculations, 23–24
 series wattage calculations, 24
fortune-telling machine. *See* Moving Message Destiny Predictor
forward bias, 22
fourteen-stage binary ripple counter, 112
frames in text animation, 241–242
frequency
 binary ripple counter, 112
 measuring with digital oscilloscope screen. *See* digital oscilloscope screen

Index

frequency of oscillation (F)
   555 timing formulas, 43
   rear LED flasher circuit, 53
front LED flasher, 56–57

G
ghost image, 163
glove LEDs, 222
gold resistor color code, 15
gray resistor color code, 15
green color
   backpack illuminator assembly code, 182
   resistor color codes, 15
"green" LED flashlight
   capacitors, 31–34
   project, 34–40
groovy light stick
   completing enclosure, 215–217
   further modifications, 222
   how circuit works, 211–212
   how to build, 213–215
   making it move, 220–222
   overview, 211
   parts, 212–213
   PIC Microcontroller program, 218–220
   stripboard layout, 213–214
   testing board, 217
grounding yourself, 12

H
hacksaw, 3
Hammond enclosure, 102
handle, light sword, 142
HDSP-5503 device, 95
hex code
   backpack illuminator, 183
   digital oscilloscope screen, 196
   dot-matrix counter, 228–229
   downloading, 14
   groovy light stick, 220
   LED light sword, 148–149
   low-res shadow camera, 208
   manually typing, 134
   Moving Message Destiny Predictor, 237–238
   PIC Microcontroller, 135, 137
   three-digit counter, 171

I
ICs (integrated circuits). See integrated circuits (ICs)
illumination
   backpack illuminator. See backpack illuminator
   LED, 22–25
   multiple LED, 25
infrared (IR) LEDs
   defined, 151
   invisible secret code display, 152–153

inputs
   BCD, 152
   color-changing light box, 64
   displaying voltage, 188
   LED light sword, 141
   LED scanner, 130
   microcontroller, 128
   mini digital display scoreboard, 75
   oscilloscope, 199
   three-digit counter, 166
insulated copper wire, 192
insulated nuts, 80
insulated screws, 80
insulation tape, 5
integrated circuits (ICs). See also circuits
   74HC range of, 85–86
   antistatic precautions, 12
   binary ripple counter. See LED binary ripple counter
   color-changing disco lights, 98
   inserting into sockets, 7
   programming PIC Microcontroller, 136–137
   seven-segment display LEDs, 73
   stripboard for, 4
invisible secret code display
   enclosure, 155–156
   future modifications, 160
   how circuit works, 152–153
   how to build and test, 156–157
   overview, 151–152
   parts, 154
   putting it all together, 158
   sending message, 158–159
   stripboard layout, 154–155

J
joint soldering, 9–12
jumper links
   in color-changing disco lights, 102–103
   in mini digital display scoreboard, 75
   replacing with mode switch, 82
   testing LED binary ripple counter, 116
   testing LED scanner, 132

K
K, 22

L
layouts, stripboard. See stripboard layouts
LDR (light-dependent resistor), 203, 210
lead-free solder, 3, 9
leads
   capacitor, 32
   cutting, 5
   dirty, 11
   groovy light stick, 214

LED illumination, 22
   in multicolor LEDs, 61–62
   secret code display, 157
   soldering, 10
   soldering bike flasher, 57
LED bike flasher
   alternative circuit, 60
   enclosure for LED bike flasher, 57–59
   experimenting to reduce current consumption, 59–60
   front LED flasher, 56–57
   how to build, 55–56
   overview, 51
   parts list, 53–54
   rear LED flasher circuit, 52–53
   stripboard layout, 54–55
LED binary ripple counter
   4060 and 74HC4060, 112
   overview, 111
   project, 112–118
LED candle. See flickering LED candle
LED flasher
   digital oscilloscope screen project and, 188
   project, 45–49
   testing digital oscilloscope screen project, 197–200
LED flashlight
   making "green," 34–40
   project, 25–30
LED light sword
   enclosure, 142–144
   final tests, 149
   how circuit works, 139–141
   how to build and test, 144–145
   LED string, 145–146
   overview, 139
   parts, 141–142
   PIC Microcontroller program, 148–149
   putting it all together, 146–148
   stripboard layout, 144
   time to play, 149
LED multiplexing
   principles, 164
   three-digit counter, 164–172
LED scanner
   how circuit works, 128–130
   how to build and test, 131–132
   parts, 130
   PIC Microcontrollers, 132, 134–138
   stripboard layout, 131
LEDs (light-emitting diodes)
   animations. See Moving Message Destiny Predictor
   color-changing light box, 63–70
   defined, 21–25
   driving multiple LEDs from single IC output. See color-changing disco lights

experimental LED sequencer circuit. *See* experimental LED sequencer circuit
light dependent. *See* experimental low-res shadow camera
light stick. *See* groovy light stick
multicolor, 61–63
seven-segment display LEDs. *See* seven-segment display LEDs
three-digit counter. *See* three-digit counter
letters. *See* characters
light dependent LEDs. *See* experimental low-res shadow camera
light-dependent resistor (LDR), 203, 210
light-emitting diodes (LEDs). *See* LEDs (light-emitting diodes)
light shield, 208
light stick. *See* groovy light stick
light sword. *See* LED light sword
light tube, 143
lighting
color-changing disco lights. *See* color-changing disco lights
color-changing light box, 70
with LEDs, 21
mcd, 26
LM3914 bargraph driver chip
digital oscilloscope screen, 188–189
vs. PIC Microcontroller, 140
LochMaster 4.0, 2, 250
low-res shadow camera. *See* experimental low-res shadow camera

M
mA (milliamps). *See* milliamps (mA)
magazines, electronics hobbyist, 250
manually operated sequencer. *See* invisible secret code display
mark-to-space ratio
555 timing formulas, 43
rear LED flasher circuit, 53
mask, camera sensor, 208–210
matrix display
backpack illuminator, 174
combinations, 177
digital oscilloscope screen. *See* digital oscilloscope screen
dot-matrix display. *See* dot-matrix display
mcd (millicandelas), 26
messages, destiny predictor. *See* moving message destiny predictor
messages, secret code, 158–159
μF (microfarads), 32
Microchip Technology Inc., 250
microcontrollers. *See* PIC Microcontrollers

microfarads (μF), 32
milliamps (mA)
74HC range of ICs, 86
4026B IC, 73
BC109 NPN current capability, 172
color-changing light box, 97
digital oscilloscope screen, 189–190
dot-matrix counter outputs, 225
limiting series resistors, 112
three-digit counter current consumption, 166
ULN2003 IC, 98
millicandelas (mcd), 26
mini digital display scoreboard
future modifications, 82
how circuit works, 73–75
how to build and test, 77–80
mounting board in enclosure, 80–82
parts, 75–76
stripboard layout, 77
mini-message display, 247
miniaturization of electronics, 210
monostable mode, 42
mounting
backpack illuminator display board, 179
backpack illuminator stripboard, 178
color-changing disco lights, 107–108
experimental low-res shadow camera, 208–210
LED flasher on bike, 57–59
LEDs on fabric, 184–186
mini digital display scoreboard, 80–82
RGB LEDs, 66
moving characters, 238–239
moving light stick, 220–222
Moving Message Destiny Predictor
construction details, 242–243
discover destiny, 245–247
display card, 243–245
final word, 247
how circuit works, 232–233
how to build and test, 234–236
overview, 231
parts, 233–234
PIC Microcontroller program, 236–242
possible modifications, 247
stripboard layout, 234–235
multicolor LEDs
backpack illuminator. *See* backpack illuminator
color-changing light box, 63–70
groovy light stick, 222
overview, 61–63
in sequencer circuit, 95–96
multimeter
fault-finding stripboard circuit, 7
testing backpack illuminator, 180

testing color-changing disco lights, 102
testing "green" LED flashlight, 37–38
testing RGB LEDs, 68
multiple LED illumination, 25
multivibrators, 42

N
nanofarads (nF), 32
NO (normally open) switch, 225
non-flammable insulation tape, 5
nonelectrolytic capacitors, 49
normally open (NO) switch, 225
Notepad, 134
numeric display LEDs
dot-matrix counter, 223–230
seven-segment display LEDs. *See* seven-segment display LEDs
three-digit counter, 164–172
nuts, insulated, 80

O
ohms (Ω)
light-dependent resistors, 203
resistor color codes, 15
series resistor calculations, 23–24
series wattage calculations, 24
orange resistor color code, 15
orienting dot matrix display, 193
oscillation frequency
555 timing formulas, 43
rear LED flasher circuit, 53
oscillators, 42
oscilloscope, 49. *See also* digital oscilloscope screen
outputs
555 timer, 42
4017 IC, 87
4026B IC, 73–75
7555 timer, 59
backpack illuminator, 174
binary ripple counter, 111–112
current consumption and light, 31
digital oscilloscope screen, 189
dot-matrix counter, 224–225
driving multiple LEDs from single IC, 87
experimenting with LED candle, 125
flickering LED candle, 120
LED illumination, 25–26
LED scanner, 129–130
low-res shadow camera, 202–203, 207
Moving Message Destiny Predictor, 232, 238
multicolor LEDs, 64–65
programming PIC Microcontroller, 135–136
rear LED flasher circuit, 52–53
source and sink currents, 44
testing low-res shadow camera, 206

P

paper diffuser sleeve, 148
parallel capacitors, 33–34
parts
   backpack illuminator, 176
   bike flasher, 54
   color-changing disco lights, 98, 100–101
   color-changing light box, 65–66
   digital oscilloscope screen, 191
   dot-matrix counter, 225–226
   "green" LED flashlight, 36
   groovy light stick, 212–213
   invisible secret code display, 154
   LED binary ripple counter, 113–114
   LED candle, 121
   LED flasher project, 46
   LED flashlight, 27
   LED light sword, 141–142
   LED scanner, 130
   low-res shadow camera, 203–204
   mini digital display scoreboard, 75–76
   Moving Message Destiny Predictor, 233–234
   three-digit counter, 167–168
PCB (printed circuit board), 1
PCLATH function, 238, 239
peddle power, 59
persistence of vision (POV)
   backpack illuminator. See backpack illuminator
   defined, 163–164
   digital oscilloscope screen. See digital oscilloscope screen
   dot-matrix counter, 223
   groovy light stick, 211
   low-res shadow camera, 201
   Moving Message Destiny Predictor, 231
   three-digit counter, 164–172
pF (picofarads), 32
PIC Microcontrollers
   backpack illuminator, 181–183
   digital oscilloscope screen, 195–196
   dot-matrix counter, 228–229
   fault-finding stripboard circuit, 7
   groovy light stick, 211, 218–220
   LED light sword, 148–149
   LED scanner, 132, 134–138
   low-res shadow camera, 207–208
   Moving Message Destiny Predictor, 236–242
   programming, 13–15
   reference books, 250
   three-digit counter, 170–171
PIC16F628-04/P microcontroller
   backpack illuminator, 174, 181–183
   digital oscilloscope screen, 189
   dot-matrix counter, 224
   LED light sword, 139
   LED scanner, 127–128
   in low-res shadow camera, 202
   Moving Message Destiny Predictor, 232
   voltage, 166
PICkit 2 Development/Programmer/Debugger, 14, 135–137
picofarads (pF), 32
pins
   555 timer, 42
   color-changing disco lights, 98, 103
   digital oscilloscope screen, 192
   dot-matrix counter configuration, 227
   experimental LED sequencer circuit, 87
   LED flasher, 45
   programming PIC Microcontroller, 135–136
   resetting LED sequencer circuit, 93
   source and sink currents, 44
   testing groovy light stick, 217
   testing LED binary ripple counter, 116
   testing LED scanner, 132
   testing low-res shadow camera, 206
   testing Moving Message Destiny Predictor, 235–236
   testing sequencer circuit, 92
plastic drilling tips, 105
polarized capacitors, 32
POV (persistence of vision). See persistence of vision (POV)
power-up button
   final tests, 149
   light sword, 140–141
PP3 clip, 180
practicing soldering techniques, 9–12
printed circuit board (PCB), 1
programming PIC Microcontroller
   backpack illuminator, 181–183
   digital oscilloscope screen, 195–196
   dot-matrix counter, 228–229
   groovy light stick, 218–220
   LED light sword, 148–149
   LED scanner, 132, 134–138
   low-res shadow camera, 207–208
   Moving Message Destiny Predictor, 236–242
   overview, 13–15
   three-digit counter, 170–171
projects
   backpack illuminator. See backpack illuminator
   basic single-LED flasher, 45–49
   color-changing disco lights. See color-changing disco lights
   color-changing light box, 63–70
   digital oscilloscope screen. See digital oscilloscope screen
   dot-matrix counter, 223–230
   experimental LED sequencer circuit. See experimental LED sequencer circuit
   flickering LED candle. See flickering LED candle
   "green" LED flashlight, 34–40
   groovy light stick. See groovy light stick
   invisible secret code display. See invisible secret code display
   LED bike flasher. See LED bike flasher
   LED binary ripple counter, 112–118
   LED flashlight, 25–30
   LED light sword. See LED light sword
   LED scanner. See LED scanner
   low-res shadow camera. See experimental low-res shadow camera
   mini digital display scoreboard. See mini digital display scoreboard
   Moving Message Destiny Predictor. See Moving Message Destiny Predictor
   three-digit counter, 164–172
protection while soldering, 3

R

R (resistance) value, 23–24
RC (resistor/capacitor) network
   binary ripple counter, 112–113
   in LED scanner, 129
rear LED flasher circuit, 52–53
rectifier diodes, 35
red color
   backpack illuminator assembly code, 182
   resistor codes, 15
Red, Green, Blue (RGB) LEDs
   color-changing light box, 63–70
   defined, 62–63
resetting
   experimental LED sequencer circuit, 93
   mini digital display scoreboard, 75, 79
resistance (R) value, 23–24
resistor/capacitor (RC) network
   binary ripple counter, 112–113
   in LED scanner, 129
resistors
   555 timing calculations, 43
   alternative circuit for bike flasher, 60
   bike flasher timing, 52
   in color-changing light box, 64–65
   color codes, 15
   digital oscilloscope screen, 189–190
   dot-matrix counter, 224–225

in experimental LED sequencer
    circuit, 88
experimenting with LED candle,
    124–125
experimenting with values, 49
experimenting with waveforms,
    198–199
groovy light stick, 211–212
invisible secret code display, 152
LED binary ripple counter,
    112–113
LED flasher project, 46–48
LED flashlight, 28
light-dependent, 203
mini digital display scoreboard,
    74–75, 78
Moving Message Destiny
    Predictor, 232
series resistor calculations, 23–24
testing digital oscilloscope screen,
    194–195
ULN2003 IC, 98
variable. *See* variable resistor
    (VR1)
resources, list of useful, 249–250
reverse bias, 22
RGB (Red, Green, Blue) LEDs
    color-changing light box, 63–70
    defined, 62–63
rotating light stick, 221
RS Components Ltd, 249

S

safety tips
    capacitors, 32
    drilling into plastic or acrylic, 105
    "green" LED flashlight, 35
    keeping stripboard dry, 184
    non-flammable insulation tape, 5
    oscilloscope input, 199
    soldering battery holder, 123
    soldering iron, 3
    using insulated screws and
        nuts, 80
    warnings, 16–17
salvaged enclosures
    bike flasher project, 56–57
    color-changing disco lights, 101
    flickering LED candle, 124
sanding stripboard, 4
scanner. *See* LED scanner
schematic diagrams. *See also* circuit
        diagrams
    converting to stripboard layouts, 2
    stripboard layouts and, 15–16
scoreboard. *See* mini digital display
        scoreboard
screws, insulated, 80
scrolling text, 241–242
secret code display. *See* invisible
        secret code display
sensor array mask, 208–210

sequencer circuits
    74HC range of ICs, 85–86
    experimental LED. *See*
        experimental LED sequencer
        circuit
    invisible secret code display. *See*
        invisible secret code display
    low-res shadow camera, 203
    three-digit counter, 166
series resistors. *See also* resistors
    4060 current consumption, 112
    alternative circuit for bike
        flasher, 60
    calculations, 23–24
    in color-changing disco lights, 103
    invisible secret code display, 152
    invisible secret code display
        modifications, 160
    in rear LED flasher circuit, 53
seven-segment display LEDs
    dot-matrix counter and, 223
    how circuit works, 73–75
    how to build and test, 77–80
    mounting board in enclosure, 80–82
    overview, 71–73
    parts, 75–76
    stripboard layout, 77
    three-digit counter, 166
shadow camera. *See* experimental
        low-res shadow camera
shaking light stick, 220
signal diodes
    in flickering LED candle, 120
    in LED scanner, 129
silver resistor color code, 15
single-digit seven segment display, 71
single in-line (SIL) sockets
    experimental LED sequencer
        circuit, 88, 90
    mini digital display scoreboard, 78
sink current
    bike flasher, 51
    defined, 44
    dot-matrix counter, 225
    experimenting to reduce current
        consumption, 59
    Moving Message Destiny
        Predictor, 232
sizes of stripboard, 1–2
sleep mode, 218
software
    downloading assembly and hex
        code, 14
    LochMaster 4.0, 2
    programming PIC Microcontroller.
        See programming PIC
        Microcontroller
solder, 8–9
solder sucker, 9
soldering
    battery holder, 123
    dot-matrix counter, 227

LED string, 145–146
LEDs for light stick, 214–215
LEDs in disco lights project,
    106–108
light sword stripboard, 143
low-res shadow camera stripboard,
    205–206
stripboard, 2
tips and techniques, 8–12
before you get started, 3
soldering iron stands, 8
soldering irons
    safety tips, 3
    temperature and angle, 10
    types of, 8
source current
    bike flasher, 51
    defined, 23, 44
specifications
    backpack illuminator, 173
    basic single-LED flasher, 45
    bike flasher, 52
    color-changing disco lights, 98
    color-changing light box, 63
    digital oscilloscope screen, 188
    dot-matrix counter, 223
    experimental LED sequencer
        circuit, 86
    "green" LED flashlight, 34
    invisible secret code display, 152
    LED binary ripple counter, 112
    LED candle, 119
    LED light sword, 139
    LED scanner, 128
    low-res shadow camera, 201
    Moving Message Destiny
        Predictor, 232
    PIC16F628-04/P microcontroller, 128
    three-digit counter, 164
speed
    4017 counter, 92
    color-changing disco lights, 110
    color-changing light box, 64, 70
    LED flasher, 47–48
    LED scanner, 138
    programming groovy light stick,
        218–219
    programming LED scanner,
        134–135
    programming light sword, 148
    programming oscilloscope, 196
sponge, 8
static electricity, 12
string, LED, 145–146
stripboard
    bike flasher project, 54
    how to build circuits on, 2–8
    incorporating display into fabric,
        184–186
    LED flasher project, 46–48
    light sword tip, 143
    preparing for solder, 9

stripboard *(continued)*
  software, 250
  working with, 1–2
stripboard layouts
  backpack illuminator, 177
  circuit diagrams and, 15–16
  color-changing disco lights, 101–102
  color-changing light box, 66
  digital oscilloscope screen, 192
  dot-matrix counter, 226–227
  "green" LED flashlight, 36–37
  groovy light stick, 213–214
  invisible secret code display, 154–155
  LED bike flasher, 54–55
  LED binary ripple counter, 114–115
  LED candle, 122
  LED flasher project, 46–47
  LED light sword, 144
  LED scanner, 131
  low-res shadow camera, 204–205
  mini digital display scoreboard, 77
  Moving Message Destiny Predictor, 234–235
  three-digit counter, 168–169
  tips, 39
strobe effect, 141
supercapacitors, 34
switches
  dot-matrix counter, 225
  final light sword tests, 149
  "green" LED flashlight, 39
  groovy light stick, 217
  invisible secret code display, 153
  LED flashlight, 26–27, 29
  LED light sword, 140–141
  LED scanner, 130, 131
  low-res shadow camera, 202
  mini digital display scoreboard, 75
  Moving Message Destiny Predictor, 233, 239
  programming LED scanner, 134–135
  programming light stick, 218
  programming oscilloscope, 196
  replacing jumper pins with mode switch, 82
  testing digital display scoreboard, 79–80
  three-digit counter, 171–172
symbols
  diodes and LEDs, 22
  multicolor LEDs, 63
  schematic diagram, 16

T
T (total time period)
  555 timing formulas, 43
  rear LED flasher circuit, 53
T1 (charge time)
  555 timing formulas, 42–43
  rear LED flasher circuit, 52
T2 (discharge time)
  555 timing formulas, 42–43
  rear LED flasher circuit, 52
temperature of soldering iron, 10
testing
  backpack illuminator, 184
  backpack illuminator boards, 180–181
  color-changing disco lights, 102–104, 107
  color-changing light box, 66, 68
  digital oscilloscope screen, 194–195
  dot-matrix counter, 227
  experimental LED sequencer circuit, 90, 92
  flickering LED candle, 123
  "green" LED flashlight, 37–39
  groovy light stick, 217
  invisible secret code display, 157
  LED binary ripple counter, 116
  LED light sword, 145, 149
  LED scanner, 132
  LED string, 146
  low-res shadow camera, 206
  mini digital display scoreboard, 78–79
  Moving Message Destiny Predictor, 235–236
  three-digit counter, 168, 170
  waveforms, 196–199
text animation, 241–242
text tables, 239
three-digit counter
  display codes, 166–167
  how circuit works, 164–166
  how to build and test, 168–170
  parts, 167–168
  PIC Microcontroller program, 170–171
  possible enhancements, 172
  POV in action, 171–172
  stripboard layout, 168–169
timer, 555. *See* 555 timer
total drive current
  defined, 44
  LED circuit considerations, 23
total time period (T)
  555 timing formulas, 43
  rear LED flasher circuit, 53
tracks, stripboard
  defined, 2
  how to cut, 3–4
transistors
  4060 binary ripple counter and, 112
  BC109 NPN, 172
  in color-changing light box, 64
  three-digit counter, 166

tricolor LEDs
  backpack illuminator, 173–174
  defined, 62
  groovy light stick, 222

U
ULN2003 IC
  74HC range of ICs, 98
  dot-matrix counter, 225
  Moving Message Destiny Predictor, 232
  seven stages of, 104

V
variable resistor (VR1)
  binary counting, 116, 118
  in color-changing disco lights, 104
  color-changing disco lights speed, 110
  experimental LED sequencer circuit, 92, 95
  LED flasher project, 45, 47
voltage
  BC109 NPN current capability, 172
  capacitor, 32
  CMOS ICs, 85–86
  digital oscilloscope screen, 190–191
  display, 188
  "green" LED flashlight, 35
  IR LED, 153
  LED binary ripple counter, 112
  LED flashlight, 26
  oscilloscope input, 199
  series resistor calculations, 23–24
  series wattage calculations, 24
  setting PICkit 2 programmer, 136–137
  three-digit counter, 166

W
warnings, 16–17. *See also* safety tips
wattage (W)
  current consumption. *See* current consumption
  LED circuit considerations, 23
  light-dependent resistors, 203
  series wattage calculations, 24–25
waveforms, digital oscilloscope screen, 196–199
wire links
  cutting, 5
  display board, 178–179
  three-digit counter, 168

X
XRATE, 242

Y
yellow resistor color code, 15